The Crucible of Creation

The Crucible of Creation

The Burgess Shale and the Rise of Animals

SIMON CONWAY MORRIS

Professor of Evolutionary Palaeobiology,
University of Cambridge

OXFORD   NEW YORK   MELBOURNE

OXFORD UNIVERSITY PRESS

1998

Oxford University Press, Great Clarendon Street, Oxford OX2 6DP

Oxford   New York

Athens   Auckland   Bangkok   Bogota   Bombay
Buenos Aires   Calcutta   Cape Town   Dar es Salaam
Delhi   Florence   Hong Kong   Istanbul   Karachi
Kuala Lumpur   Madras   Madrid   Melbourne
Mexico City   Nairobi   Paris   Singapore
Taipei   Tokyo   Toronto   Warsaw
and associated companies in
Berlin   Ibadan

Oxford is a trade mark of Oxford University Press

Published in the United States
by Oxford University Press, Inc., New York

© Simon Conway Morris, 1998

A catalogue record for this book is available from the British Library

Library of Congress Cataloging in Publication Data
(Data applied for)

ISBN 0 19 850256 7

Typeset by EXPO Holdings, Malaysia

Printed in Great Britain
by
Bookcraft Ltd., Midsomer Norton, Somerset

To Andrew and Hugh

Preface

To the best of my knowledge this is the fourth book to be published that considers as its main theme the Burgess Shale and its half-billion-year-old fauna. Is not the market place becoming a little crowded? When we consider the works already available I hope that I may persuade you, at least by the time you have finished reading this book, that not only are there new viewpoints, but by no means the last word has been written. The first book to appear (*The Burgess Shale*) was issued in 1985 by the leader of the Cambridge team, Harry Whittington, who was also my research supervisor. This book is straightforward and provides a concise, readable description. The most recent book (1994) is *The fossils of the Burgess Shale*, whose senior author, Derek Briggs, was also supervised by Harry Whittington. His co-authors Doug Erwin and Fred Collier are respectively a palaeobiologist based in the Smithsonian Institution with wide interests in evolution and the Cambrian radiations, and a former Collections Manager of the same Institution, who is now based in the Museum of Comparative Zoology, Harvard University. In passing, I should say that when we started our research projects Fred Collier was instrumental in allowing Derek and me free and unrestricted access to the Burgess Shale collections in the Smithsonian Institution. Their book is a guide to the Burgess Shale, well illustrated and providing an overview of the fauna and flora. In many respects it is an updated and more polished version of *The atlas of the Burgess Shale*, which I edited and was published as a laboratory handbook by the Palaeontological Association in 1982. The book by Derek Briggs and his colleagues makes no real attempt to assess the evolutionary importance and controversies that have grown up around the Burgess Shale. It is with the third book, by Stephen Jay Gould (*Wonderful life*), that some of the general implications of the Burgess Shale were drawn to the attention of a wider public. One of my purposes in this book is to discuss the basis upon which Gould builds many of his conclusions, especially concerning the roles of historical contingency and the evolutionary explanations for the apparently remarkable range of morphological types we see in the Cambrian. In *Wonderful life* Gould is kind enough to say that one part of his inspiration for his book came from some short remarks I had made about one Burgess Shale animal, *Wiwaxia*, in a scientific paper published in 1985. Gould is also very generous in his praise for our research work on the Burgess Shale, and I must certainly acknowledge that his writings have done much to sharpen my appreciation of the issues involved. As will be clear from this book I believe, however, that a different message can be read from the Burgess Shale than that promoted by Gould. In part this reorientation stems from important new palaeontological information that has emerged, notably concerning *Hallucigenia* and *Anomalocaris*. Linked to this are new discoveries, especially from South China and North Greenland, that seem to throw remarkable new light on the origin of evolutionary novelties and the emergence of body plans. Such discoveries appear to cast considerable doubt on various mechanisms of macroevolution, including some briefly discussed in *Wonderful life*. Other issues, however, are at heart philosophical, and a resolution of differences need not rely solely on the scientific data. As I argue with particular respect to the role of contingency, it is not so much that such a factor is important—indeed at one level it is—but whether it has any meaningful effect on the totality of life, given the constraints of existence and the ubiquity of evolutionary convergence.

To assist those who might be interested in exploring the scientific literature, each chapter is accompanied by a set of short notes. These provide some guidance to further reading and, in places, a limited commentary. It is emphatically not my intention to provide an exhaustive review of the relevant literature, and the many omissions should not be viewed in a sinister light. Let me conclude by saying that although this book is primarily aimed at documenting a particular episode of scientific investigation in which I had the good fortune to be involved, any success it enjoys will be because of some of the wider issues raised. If science faces a particular challenge in the next few years, I believe it will not be from hewing away at the frontiers but from the necessity to reconcile its conclusions with other truths, many of which carry a particular ethical urgency.

Cambridge   SCM
November 1997

Acknowledgements

My primary debt is to Harry Whittington, who first allowed me to join the Burgess Shale project and since then has provided support, inspiration, and friendship. I owe similar thanks to John Peel, who also introduced me to the high Arctic of North Greenland. Many other individuals have offered unstinting help and assistance and I would like to mention especially Jim Aitken, Stefan Bengtson, Derek Briggs, Graham Budd, Nick Butterfield, Chen Junyuan, Fred Collier, Des Collins, Mary Droser, Bernie Erdtmann, Yoshi Endo, Doug Erwin, Bill Fritz, the Gunther family, Peter Jell, Richard Jenkins, Andy Knoll, Ed Landing, Ian McIlreath, Ken McNamara, Dick Robison, Bruce Runnegar, Shu Degan, Alberto Simonetta, Jim Valentine, and Ellis Yochelson. I should also like to give special thanks to Sandra Last for heroic typing and organization, and my colleagues in Cambridge, especially Liz Harper and Rachel Wood. Liz Harper and Harry Whittington kindly read the entire draft manuscript, and John Peel reviewed Chapter 4. I am very grateful for their many and helpful suggestions. In addition to the various individuals who generously made available illustrations which are acknowledged separately, several people also provided other specific help. Dr A. Casinos (University of Barcelona) let me have unpublished data on the weight of the giant sloth. Professor D.L. Bruton (University of Oslo) generously made available unpublished notes and letters of his season in the Walcott Quarry in 1967. Dr P. Janvier (Museum of Natural History, Paris) provided a detailed and fascinating account of Henri Mansuy. Technical help by Dudley Simons and Hilary Alberti is also acknowledged. The department in Cambridge has been and remains very supportive, and I especially value their commitment to long-term results as against short-term expediency. None of this work would have been possible without the long slog of collecting. In addition to being given open access to the vast Walcott collections, I must warmly thank the collecting teams operating for the Geological Survey of Canada and the Royal Ontario Museum, and also the colleagues and assistants on our Greenland expeditions (Mona Bjørklund, Jon Ineson, Paul Schiøler, and Paul Smith). My research has been supported by the Natural Environment Research Council, the Carlsberg Foundation, National Geographic, the Cowper Reed Fund, and St John's College, Cambridge. Finally, I wish to thank the staff of Oxford University Press for their encouragement and humour during this enterprise.

Contents

Glossary

Asterisks refer to other entries in the glossary.

acritarchs. Organic-walled microfossils, generally interpreted as the reproductive cysts of algae. The taxonomic relationships of acritarchs are generally controversial.

amino acid. The basic building-blocks of proteins, composed of the amino group (NH_2), the carboxyl group (COOH), a hydrogen atom and the side chain attached to the carbon atom. About 20 amino acids are routinely used by living organisms.

annelid. A phylum* of metazoans.* The body consists of a head and a series of segments.* Each segment bears bristle-like chaetae* composed of chitin.* The internal anatomy includes a spacious fluid-filled body cavity (coelom).

anthozoans. A group of cnidarians.* Typical examples include the sea anemones, corals, and pennatulaceans.* The body forms a so-called polyp, with tentacles equipped with stinging cells surrounding the mouth.

arthropods. A phylum of metazoans.* The body is segmented and has an exoskeleton* divided into jointed units. The appendages are often specialized for different functions, especially in the vicinity of the mouth where they handle and crush food.

bacteria. Micro-organisms that show an enormous range of metabolic strategies and may inhabit extreme environments. The cells have a prokaryotic organization, lacking a central nucleus and organelles. Two major groups of bacteria are recognized: the eubacteria, e.g. cyanobacteria,* and archaebacteria, e.g. eocytes.*

benthos. Those organisms that live on or in the substrate, typically in marine sediments. There is a further distinction between organisms living on the sediment surface, the epifauna* or epiflora, and those living in the sediment as infauna.* The marine benthic zone is inhabited from the surf zone to the deepest part of the oceans, but the epiflora of algae and seaweeds is restricted to the photic zone.*

brachiopods. A phylum of metazoans.* The body is enclosed within two shells, usually composed of calcium carbonate, but sometimes of calcium phosphate. The soft parts include a complex feeding apparatus (the lophophore) with tentacles that are used for suspension-feeding.* Around the margins of each shell there is a fringe of setae composed of chitin* and that have the same microstructure of the chaetae* of polychaetes.*

Burgess Shale. A sedimentary unit forming part of the Stephen Formation. It is Middle Cambrian in age, *c.* 520 million years old. The Burgess Shale is located in the Rocky Mountains of British Columbia, Canada. Renowned for its superbly preserved soft-bodied fossils.

cataphract. An arrangement of sclerites* reminiscent of chain-mail, whereby they are arranged in a closely imbricated fashion that covers the surface of the body.

Cathedral Escarpment. A cliff-like structure that in Burgess Shale* times separated the shallow waters of the continental shelf from the deeper waters where the soft-bodied animals flourished. As such it is not directly visible today, at least in its entirety. Rather it is recognized and reconstructed from the stratigraphic sequences in the region of the Burgess Shale, and notably from study of a unit known as the Cathedral Formation. This takes its name from the Cathedral Crags, spectacular pinnacles of rock adjacent to Mount Stephen.

chaetae. Bristle-like structures that are composed of chitin* and arise on either side of each segment* of an annelid* worm. Identical structures are found in brachiopods,* where they are referred to as setae. Both chaetae and setae are believed to have derived from the sclerites of halkieriids.*

chaetognaths. A phylum of metazoans,* sometimes known as arrow worms. Most species are active swimmers, with prominent fins and at the anterior end a feeding apparatus composed of prominent teeth that grasp prey. Their relationships have been controversial, but evidence from molecular biology indicates that chaetognaths belong within the protostomes.*

Chengjiang. A Lower Cambrian deposit, about 530 million years old, located in Yunnan Province, South China. It is famous for its soft-bodied fossils, similar to those of the Burgess Shale.*

chitin. A substance, widely employed as structural component of animals, such as the exoskeletons* of arthropods* and the chaetae* of annelids.* Its molecular structure is similar to that of a polysaccaharide. Chitin is often strengthened by a chemical process of tanning.

chordates. A phylum of metazoans.* The body is segmented, and in the primitive state the muscles form characteristic structures known as myotomes.* The other characteristic feature is a stiff rod running along much of the body, known as the notochord.* Chordates encompass all vertebrates, including Man, and the more primitive cephalochordates (amphioxus).

cilia. Minute hair-like structures that arise singly or in combination from the cells of many eukaryotes. The internal structure of microfibrils is very characteristic. Cilia are employed for various functions, including suspension-feeding.*

clade. A set of species that share a common ancestor, and thus form some part or branch of the evolutionary tree.

cnidarians. A phylum of metazoans.* Divided into four main groups, one of which is the anthozoans.* The body consists of three basic layers: inner endoderm, middle mesoglea*, and outer ectoderm. Cnidarians take their name from the characteristic stinging cells known as cnidae or nematocysts.

contingency. The notion that the present-day world arises as a result of chance events in the past, some of which were in themselves trivial and at the time seemingly of minor importance.

coxa. The unit of an arthropod* appendage that gives rise to the gill and walking leg. In many arthropods the coxa bears spines or other projections that are used for feeding.

craton. The nucleus of a continent, typically about 25 km in thickness and composed largely of granitic rocks. Cratons are usually relatively stable and seismically inactive.

ctenophores. A phylum of metazoans.* Most species are planktonic* and have a gelatinous body that bears eight rows of cilia known as the comb-rows. The beating of these cilia* help to propel the ctenophore through the water. Ctenophores are active predators, and many capture prey by using long tentacles equipped with sticky cells.

cultrate. A type of sclerite, found in the halkieriids,* which is usually symmetrical in form.

cyanobacteria. A group of eubacteria, sometimes known as the blue-green algae. This group is photosynthetic, and forms microbial mats that may build stromatolites.*

deuterostomes. A superphylum that encompasses the chordates,* hemichordates,* and echinoderms.* Deuterostome means 'second mouth' and refers to the fact that the mouth is secondarily formed in the embryo, in contrast to the protostomes.*

DNA. Deoxyribonucleic acid, the basis of the genetic code. The four nucleotides (adenine, cytosine, guanine, thymine) are arranged in a double helix. Some DNA codes for particular products, but typically much of the DNA in eukaryotes has no clear function and may be 'junk'.

echinoderms. A phylum of metazoans,* with a characteristic five-fold symmetry. Typical representatives include echinoids (sea urchins) and asteroids (starfish). The anatomy includes the diagnostic water-vascular system and the calcareous skeleton has a unique porous structure (stereom).

Ediacaran fauna. Fossils of latest Proterozoic age, *c.* 560 million years old. The species are almost entirely soft-bodied, and most seem to be related to the cnidarians.* Possible examples of primitive protostomes* and conceivably deuterostomes* are also known. Ediacaran fossils are still controversial, and some scientists have suggested they represent a new group known as the vendobionts.

eocytes. A group of bacteria,* specifically archaebacteria. Many eocytes live in extreme environments, including hot springs. Some molecular information suggests eocytes are more closely related to the eukaryotes than any other group of bacteria.

epifauna. Those animals that live on the sea floor, in contrast to those that live within the sediment as infauna* or above the seabed in the pelagic* realm. Epifauna may be either attached to the sediment surface (sessile*) or roam about the sea floor (vagrant).

exoskeleton. The external skeleton, most typical of the arthropods.* In this phylum the basic composition of the exoskeleton is chitin,* but in groups such as the trilobites* it is reinforced by minerals such as calcium carbonate.

genome. The totality of DNA,* housed on chromosomes, in a cell, including all sequences that are potentially or actually transcribed (exons) and other intervals that are either discarded after transcription (introns) or have no known function and in some cases may be 'junk'.

geological time. Objective geological time is measured by radiometric methods that assume (reasonably) that decay constants of radioactive isotopes remain unchanged. Many systems exist, including those comparing ratios of radiogenic argon ($^{40}Ar/^{39}Ar$), potassium and argon (K/Ar), rubidium and strontium (Rb/Sr), and uranium and lead (U/Pb). The last of these methods is providing some very precise determinations, but any radiometric method has inherent errors that are seldom less than a million years. Geological time can also be assessed by the accumulation of rocks in stratigraphic sections and their correlations to rocks in other parts of the world. Most correlations rely on fossils (biostratigraphy), but a number of other methods (e.g. chemostratigraphy, magnetostratigraphy) are also employed.

haemocoel. A body cavity that is filled with blood, and is especially characteristic of the arthropods.*

halkieriids. An extinct group of Cambrian invertebrates. The skeleton is composed of numerous sclerites* that form an articulated array over the upper surface of the body. In addition, at least some species carried a prominent shell at either end of the body. Halkieriids are believed to throw significant light on the origins of several phyla including the annelids,* brachiopods,* and molluscs.*

hemichordate. A phylum of animals that belong to the deuterostomes.* Living examples include the acorn worms and rhabdopleurids. The latter group is widely accepted as being closely related to the extinct graptolites.

holothurians. A group of echinoderms,* also known as the sea cucumbers. The skeleton is reduced to isolated spicules embedded in a leathery skin, but other characteristic features of the echinoderms such as a fivefold symmetry and a water vascular system are always present. Holothurians live in all parts of the sea. Bizarre types have been discovered in the deep sea. Those from shallow waters are a popular delicacy (Bêche-de-mer) in many parts of the world.

Hox **genes.** A specific class of genes that play an important role in animal development, especially the demarcation of major body divisions during embryology. *Hox* genes are very widespread in metazoans,* and may be a defining character of this kingdom. These genes have a characteristic motif that expresses a protein involved with DNA* binding.

hyoliths. An extinct group, generally either placed in the molluscs* or close to this phylum.* The external skeleton is calcareous and composed of several units. Most of the soft parts are housed in an elongate cone, which could be closed off by a lid-like operculum. Projecting on either side of the operculum was a pair of narrow recurved struts, known as the helens. These are believed to have been used for locomotion.

infauna. Those animals that live within the sediment of the seabed. As with the epifauna,* both sessile* and vagrant modes of life are known. Many infaunal animals make burrows, which may be preserved as trace fossils.*

Lagerstätte. A fossil deposit that provides an exceptional insight into the former diversity of life. Broadly two types of *Lagerstätten* are recognized. The first are Conservation-Lagerstätten, in which soft-part preservation (such as in the Burgess Shale*) is common. The other type is known as the Concentration-Lagerstätten, and are represented by such deposits as bone beds and fissure fills of caves.

Laurentia. The major craton,* identifiable at least as far back as the late Proterozoic, which effectively comprises the United States, Canada, and Greenland. Burgess Shale-type faunas show an approximately concentric distribution around the Laurentian craton.

lobopod. The soft walking appendage of a lobopodian, a group of primitive arthropods* that flourished in the Cambrian, e.g. *Hallucigenia*, *Microdictyon*, and are represented today by the onychophores.* The fluid-filled cavity within the lobopod is the haemocoel.*

lophophores. The tentacular feeding organ of the brachiopods* and their near-relatives the ectoprocts (or bryozoans) and phoronids. The lophophore is employed in suspension-feeding.* A closely similar structure is found in the hemichordates*, a phylum* of deuterostomes.* This similarity is probably due to convergent evolution because the brachiopods and their relatives are now regarded as protostomes.*

mesoglea. A gelatinous layer that separates the inner (endoderm) and outer (ectoderm) layers of tissue in the Phylum Cnidaria.* In animals such as the jellyfish (Class Scyphozoa) the mesoglea is usually massively developed, and although largely composed of water also contains abundant fibrils of collagen that impart considerable strength to the body. In the ctenophores* there is also a gelatinous-like mesoglea, but it differs from that of the cnidarians by containing muscle fibres.

metazoans. One of the six kingdoms of life. Metazoans (or animals) are characterized by multicellularity, the formation of tissues including a nervous system and muscles, and an alimentary canal that may or may not have an anus. Metazoans are eukaryotes, and probably evolved from within the Kingdom Protista*, or possibly from primitive examples of the Kingdom Fungi.

molluscs. A phylum* of metazoans.* They are characterized by a broad foot used for locomotion, a dorsal shell with a calcareous composition, and a distinctive feeding apparatus consisting of a ribbon of teeth that is known as the radula. Molluscs show a considerable diversity of form. In addition to the primitive aplacophorans and monoplacophorans,* more advanced types include the snails (gastropods), clams (bivalves), and squid (cephalopods).

monoplacophorans. A primitive group of molluscs.* The dorsal shell usually forms a broad cone, and the underlying foot is more or less circular in outline. Superficially, monoplacophorans look rather similar to the limpets, which are a group of gastropods, but evidence of segmentation* indicates the retention of primitive characters.

myotomes. The muscular blocks seen in the more primitive chordates,* notably the fish and the cephalochordate amphioxus. The myotomes have a rather complex cone-in-cone structure, and on the sides of the animal appear in a characteristic zig-zag configuration. Contraction of the myotomes leads to sideways flexure of the body and hence propulsion of the animal. The notochord* forms the antagonistic organ against which the myotomes operate.

nekton. Those animals that swim actively in the water column and thus belong to the pelagic* realm. Typical members of the nekton include the fish, cephalopods such as the squid, marine mammals, and the chaetognaths.*

neurochaetae. The chitinous bristles or chaetae* that arise from the neuropodium.*

neuropodium. The lower (ventral) lobe of the parapodium* of polychaete annelids.* The neuropodium bears the neurochaetae.*

niche. An elastic and imprecise concept widely used in ecology, to refer to the way in which an organism occupies and utilizes some part of the surrounding environment, most obviously in terms of feeding, protection, and reproduction. Niche subdivision refers to the apportioning of a niche, which can operate on a variety of scales, for example, on a tree one could refer to bark and canopy niches, or to opposite sides of a leaf.

notochaetae. The chitinous bristles or chaetae* that arise from the notopodium.*

notochord. The stiffening rod found in chordates,* which in the more advanced groups provides the basis of the vertebral column. The notochord is an elastic structure that acts as an antagonist to the contraction of the muscular myotomes.*

notopodium. The upper (dorsal) lobe of the parapodium* of polychaete annelids.* The notopodium bears the notochaetae.*

onychophorans. A primitive group of arthropods.* Onychophorans lacked the jointed exoskeleton* characteristic of other members of the phylum. The body is relatively soft and flexible, and the animals walk on lobopods. Living onychophorans are terrestrial, but a wide range of marine examples are known from the Cambrian, especially the Burgess Shale* and Chengjiang.* Particular examples include *Aysheaia* and *Hallucigenia*.

palaeontologist. An individual who studies fossils. One might distinguish further palaeobotanists (plants), palaeozoologists (animals), and those more interested in the biology of extinct organisms (palaeobiologists).

palmate. A type of sclerite found in halkieriids,* which is usually somewhat asymmetrical and flattened.

parapodium. The lobe-like extension of the body that is characteristic of the polychaete annelids,* and is usually divided into the dorsal notopodium* and ventral neuropodium.*

pelagic. Those organisms that either swim (nekton*) or float (plankton*) in the water column.

pennatulaceans. A group of anthozoans* (phylum Cnidaria*), often known as sea pens. All pennatulaceans are members of the sessile epifauna,* and show a wide variety of morphologies. Most typical, perhaps, are the frond-like species with a series of branches arising from a central stem. Pennatulaceans are colonial animals, but the individual polyps are highly integrated, and the sea pen thus behaves more like a single organism.

phenotype. The expressed morphology or behaviour of an individual, the ultimate basis for which is genetic and thus resides in the genotype.

photic zone. The depth of water into which sunlight can penetrate. In clear oceanic water the photic zone may be deeper than 100 m, but in turbid waters around coasts the photic zone is much less. The zone controls the distribution of marine algae and other plants that require sunlight for photosynthesis.

phylogeny. The study of evolution, specifically the interrelationships of organisms and the branching patterns of descent from a common ancestor. Various methods of phylogenetic analysis exist, but the most popular procedure involves cladistics.

phylum. A major grouping in a scheme of biological classification, ranking next below a kingdom. Commonly included in the concept of a phylum is the notion of the body plan, with the implication that all component species are descended from a single remote ancestral form. In reality, a phylum is little more than a statement of taxonomic ignorance, and its wider relationships to other phyla is usually a matter of controversy. About 35 living phyla are recognized. They include such examples as the annelids,* arthropods,* brachiopods,* cnidarians,* and molluscs.*

plankton. Those organisms that float in the water column. Many are very small (e.g. planktonic foraminifera) but larger ones will sink very quickly unless they employ flotation devices, e.g. the gas-filled float of the Portuguese Man-o'-War (Cnidaria*).

plate tectonics. The recognition that the outer layer of the Earth is divided into a number of rigid plates that are generated along oceanic ridges by intrusion of basaltic magma. Typically this is a submarine process and the generated sea floor moves away from the spreading ridge until ultimately it descends in a subduction zone. The cratons* effectively 'float' on the tops of the plates, and although they may collide they are generally too light to be subducted.

plesiomorphy. A term widely employed in cladistic studies, which refers to a character state that is possessed by all taxa* under consideration and thus is of no help in identifying which of them are the more closely related as sister-taxa. Characters that are not plesiomorphic are known either as synapomorphic (shared between sister taxa) or autapomorphic (restricted to a taxon). In humans for example our vertebral column is plesiomorphic with snakes and birds, our hair is synapomorphic with dogs and squirrels, and our language with grammar is autapomorphic.

polychaetes. A group within the annelids.* Nearly all polychaetes are marine. Many are infaunal,* but others stroll across the seabed, and some even swim. Lateral extensions, known as parapodia,* are usually prominent and bear the bristle-like chaetae.* The head region is usually well developed and commonly bears various tentacles and sometimes eyes.

priapulids. A phylum* of marine metazoans*, sometimes referred to as the penis-worms. The body is divided into two main sections, a spiny proboscis at the anterior end and an annulated trunk. The larger living priapulids are voracious predators, but microscopic species probably live on bacteria. Priapulids are abundant and diverse in the Cambrian, notably the Burgess Shale.*

problematica. Organisms of uncertain systematic position, sometimes known as *incertae sedis*, that cannot be accommodated in known groups. Especially abundant in the Cambrian, problematic fossils continue to pose major problems in phylogenetic analysis.

proboscis. An anterior structure found in a variety of phyla, including the annelids (polychaetes*) and priapulids.* The proboscis has a variety of functions, but it is usually involved with feeding and in the priapulids it is also involved in the burrowing cycle.

protistan. The kingdom of the eucaryotes that for the most part are single-celled e.g. ciliates, *Amoeba*, but also include the multicellular seaweeds. About twenty distinct groups of protistan are recognized.

protostomes. A superphylum that traditionally has encompassed phyla* such as the annelids,* arthropods,* and molluscs.* More recently it has become clear that various other groups, including the brachiopods,* chaetognaths* and probably the priapulids* also belong within the protostomes. In contrast to the deuterostomes,* the mouth is primary, forming from the embryonic pore that forms as a result of the infolding of the embryo in a process known as gastrulation.

sclerite. A plate-like structure that is individually inserted on to the body of a metazoan* and in combination with adjacent sclerites forms a cataphract* skeleton. Sclerites are typical of the extinct halkieriids* where three distinct shapes (cultrate,* palmate,* and siculate*) are recognized.

segments. Serial divisions of the body. In annelids,* arthropods,* and chordates* the segments are referred to as metameric, and in the primitive state showed almost identical repetition throughout substantial lengths of the body. Metameric segmentation is most clearly expressed in the annelids, whereas in the arthropods and chordates series of segments tend to become specialized for particular functions. This is known as tagmosis.

sessile. This refers to animals that are either rooted, cemented, or otherwise attached in the sea floor or to any object such as seaweed or another animal.

siculate. A type of sclerite found in halkieriids,* which is recurved and occurs in characteristic bundles arising from a common base.

Sirius Passet. A locality in Peary Land, North Greenland with an abundance of soft-bodied fossils. The Sirius Passet fauna is of Lower Cambrian age, *c*. 530 million years old, and it is about the same age as the Chengjiang* fauna. Articulated halkieriids* and primitive arthropods are a noteworthy feature of the Sirius Passet fauna.

skeleton. A supportive structure that provides mechanical strength and/or protection to an organism. Although usually considered in the context of mineralized hard parts, such as the exoskeleton* of an arthropod* or the shell of a brachiopod,* a skeleton can also depend on internal fluid pressure in a hydraulic system such as occurs in the burrowing cycle of priapulids.*

species. The fundamental taxonomic unit, often recognized on the basis of anatomical features and/or behaviour, but usually defined on the basis of reproductive isolation and the ability to interbreed and produce fertile offspring.

sponges. Generally accepted as the most primitive phylum* within the meta-zoans.* Sponges are invariably sessile* and are suspension-feeders.* Although the body is organized and has a skeleton* of spicules, sponges lack organized tissues and have no nervous system.

stromatolites. Laminated structures built by a combination of microbial activity and accumulation in the form of trapping or precipitation of sediments. The top surface of a living stromatolite consists of a stratified microbial mat, the uppermost layer of which is rich in cyanobacteria,* which are photosynthesizers. Beneath this top layer a variety of other microbial groups flourish, including green and purple bacteria.*

suspension-feeder. Organisms that strain the sea water for minute suspended food particles. Most often the feeding net, such as the lophophore* of brachiopods,* is composed of a tentacular organ with abundant cilia.* In suspension-feeding arthropods* the sea water is strained by limbs modified as rakes and filters. Suspension-feeders usually have a transport system to move captured particles to the mouth for swallowing and subsequent digestion.

taphonomy. The post-mortem history of a fossil, that in principle entails the study of its pre-burial history, e.g. transport and disarticulation, and its post-burial history in the sediments, e.g. compaction.

taxon. The units within the taxonomic hierarchy,* from the fundamental unit of species* to phylum,* via a set of intermediate categories.

taxonomic hierarchy. The concept of a nested set of levels that span the divisions between the species,* widely taken to be the basic unit of evolution, and the phylum.* The most widely used divisions in between are, in ascending order, the genus, family, order, and class. The taxonomic hierarchy is widely employed in systematic biology, but it is not free from controversy. In particular it is widely agreed that with the exception of the species (and even here there are dissenters), the remaining levels are rather arbitrary constructs of human devising that only imprecisely reflect the evolutionary relationships of the organisms concerned. In addition, it is widely agreed that it is almost impossible to establish a precise equivalence between a particular taxonomic level, for example a family of fish with a family of squid.

trace fossil. The remains of the activity of an animal, exemplified by the burrow of a worm or the walking trail of an arthropod.* Trace fossils typically involve reworking of the sediment, and thus in normal circumstances of sedimentation they cannot be transported without being destroyed.

trilobites. An extinct group of arthropods* that flourished in the early Palaeozoic and eventually became extinct about 250 million years ago at the end of the Permian. Trilobites have a calcareous exoskeleton* with threefold division of the body into head shield (cephalon), mid-section (thorax), and tail (pygidium). The term *tri*-lobite, however, refers to the longitudinal division of a central lobe flanked on either side by lateral lobes.

vertebrates. The group within the chordates,* that encompasses the fish, amphibians, reptiles, birds, and mammals. All are characterized by a vertebral column, and in the tetrapods limbs variously modified for walking, swimming, or flying.

CHAPTER 1

The imprint of evolution

Introduction

We live on a wonderful planet that not only teems with life but shows a mar-vellous exuberance of form and variety. In comparison with the size of the Earth its living skin (the so-called biosphere) may be thin, but it is by no means negligible. From high in the atmosphere, where ballooning spiders wafted aloft on their silk-strings have been trapped at heights of more than 4500 m and birds such as condors cross tropical storms at altitudes well above 6000 m, via the oceans and the green continents, to deep within the Earth's crust where bacteria are known to live at depths of at least several kilometres, life is pervasive.

Nobody knows the precise total of species that presently inhabit the Earth, nor how many once existed but are now extinct. There could quite easily be twenty million species alive today, and the number of extinct species must run into the hundreds of millions, if not the billions. Within this vast pleni-tude it is perhaps rather surprising that there is only one, unique species that can understand a single word of this book. This species, which is of course ourselves, is uniquely privileged: not only can we understand something of our origins, but we are the first animals ever to have looked at the stars and seen anything more than distant pin-pricks of light.

Because, in some ways, we are utterly different from any other form of life that has ever evolved, how do we know that our origins and history are to be traced here on Earth rather than as extraterrestrial immigrants? The reason is simple: our evolutionary pedigree is stamped on every feature and permeates the entire fabric of our bodies. Some aspects of our history are of com-parative recency. For example, our ability to walk upright (the bipedal stance) was achieved only about four and a half million years (Ma) ago. The astonishing increase in our brain size, even in comparison with the closely related apes, is yet more recent. The basic structure of our arms and our legs, including the characteristic five fingers and toes (technically the pentadactyl limb), can be traced back over hundreds of millions of years. Indeed, it is now possible to study fossils, including some collected from Devonian rocks (about 370 Ma old) in east Greenland, that indicate how the fins of aquatic fish were transformed into the limbs of the first terrestrial vertebrates.[1] Similarly, although our brains are unique in their mental and spiritual facul-ties, the basic structure of the brain is easily identifiable in primitive fish. This arrangement must have evolved at least 500 Ma ago. But our evolutionary history is much more deeply encoded than organs such as limbs and brains. In many ways our basic biochemistry is little different from that of the bacteria. These steps in evolution were achieved thousands of millions of years ago. Not only do we and bacteria both use DNA for replication, but special proteins (the histones) that surround the strand of DNA and assist with keeping it stable and in the correct configuration are very similar in their sequence of building blocks (the amino acids) in all life. This is simply because they play a fundamental role in maintaining the proper function of the DNA; most alterations are automatically fatal.

It is, however, self-evident, even if the histone proteins are almost invariant in their structure, that life itself has not remained at the level of bacteria. The world is full, not only of bacteria, but also of animals as different as cranes, whales, oysters, and sharks, not to mention the plants, fungi, and single-celled organisms such as *Amoeba*. This book is not directly concerned with the origins of any of these creatures, or indeed ourselves. Rather it is an exploration of how a single unit of rock, from the west of Canada and known as the Burgess Shale, has placed the history of life, and so by implication Man's place in the scheme of evolution, in a new set of contexts.

What then is this Burgess Shale and why is it regarded as so important? How it was discovered, who worked on it, what scientific mistakes they inevitably made, how much remains to be learnt, and whether the whole concept of evolution in the Darwinian framework now needs to be radically reconsidered will all be considered in the rest of this book. The Burgess Shale is a thin unit of rock. The outcrop itself, in a small quarry on the side of a hill, is rather drab and unremarkable, but any palaeontologist would want to work there for two reasons. One is seemingly trivial: even if the quarry looks very ordinary, the Burgess Shale occurs in some of the most beautiful scenery in the world, in the Main Ranges of the Canadian Rocky Mountains. Looking from the quarry, as far as the eye can see, there are snow-capped mountains, glaciers, turquoise-coloured lakes, and forests set in wilderness. If one has to collect fossils, one might as well collect them here! The second reason is that the Burgess Shale is no ordinary fossil deposit. Here, by as yet largely unknown mechanisms,[2] the processes of rotting and decay have been largely held in abeyance so that the true richness of ancient life is revealed: not only are there animals such as trilobites and molluscs with tough, durable skeletons, but completely soft-bodied animals are also preserved. These remarkable fossils reveal not only their outlines but sometimes even internal organs such as the intestine or muscles.

The Burgess Shale is not unique, but for those who study evolution and fossils it has become something of an icon. It provides a reference point and a benchmark, a point of common discussion and an issue of universal scientific interest. Just as Darwin's finches from the Galápagos Island exemplify the recognition of the central role of adaptation, or the laboratory fly *Drosophila* stands as a symbol for the profound successes of molecular biology, so the Burgess Shale is becoming the icon for those who study the history of life. But before we begin to understand what the riches of the Burgess Shale mean, both to evolution and the scientific method, it is essential to place it in a wider context. By obtaining a sense of its place in the unfolding drama of life, set in an ecological theatre, so we can understand why it has become one of the leading players.

Evolution: why no consensus?

All science is embedded in a framework, which provides the points of reference and a necessary stability to our enterprise. Not surprisingly, many aspects of the framework remain little changed for decades, and on a day-to-day basis are accepted and usually remain unchallenged. For biology it has been famously observed that nothing makes sense unless considered in the context of evolution. The fact of organic evolution in itself is not in dispute. This is because in essence the Darwinian formulation of descent through time and co-occurring modification of the organisms, usually registered in the fossil record by anatomical changes, seems to be unanswerably correct. Once there were only bacteria; now they share the planet with millions of other types of life. Separate and special creation of each and every species is a logical alternative, and in itself need not be beyond reason. Nevertheless, the study of comparative anatomy, behaviour, molecular biology, and the fossil record give no support to any such model of recurrent creation.

So if we accept a tree of life, arising from a single ancestor approximately four thousand million years ago, why does the apparently simple fact of organic evolution excite continuing debate and disagreement? What is it that is in dispute? At heart there are two areas of contention: those of mechanism and those of implication. The first is a scientific problem, the second is meta-physical. Our immediate concern here is with the aspects of evolutionary theory (as presently portrayed) that are relevant to the Burgess Shale. As with most areas of science, the argument proceeds by reference to examples. The story of the Burgess Shale therefore epitomizes many aspects of the debate on evolution, but this extraordinary fauna is nevertheless no more than a convenient vehicle that embodies the wider principles that are at stake.

A recurrent difficulty in discussions on organic evolution is that schools of thought are too often polarized, although this is understandable because of the need to solve tractable problems that need to be stated in high circumscribed language. Nevertheless, in all the debates and disagreement, it seems rather extraordinary that for the most part it is almost forgotten that evolution is a historical process. In part it is accessible from the fossil record, and the analogies with the study of human history are clear. For example, if I wish to know more about the history of a college in Cambridge I can spend a rewarding time in the archives, aware that not all documents are decipherable and some may have been lost by fire, flood, or worm. Much can also be learnt, however, from simply studying the present order, be it of the buildings or the nature of its society. Here, too, there will be a clear historical stamp. It is all the more remarkable that the pitfalls and fallacies that the well-trained historian teaches others to avoid do not seem to be utilized by those investigating the parallels in the history of life. In a society stricken by post-Saussurean relativism[3] it is also too often forgotten that history had a unique course, and that in principle it is knowable.

What then of evolutionary mechanism? In brief, there seem to be three main problems to consider. It is widely, although not universally, agreed that central to the evolutionary process is the splitting of lineages, with at least one of the descendant forms differing materially from the ancestral type. Most biologists identify this process as one of speciation, the formation of new species. In classical biology this aspect of evolution has been construed in terms of mechanisms that promote the isolation of groups of individuals (populations) and thereby, at some subsequent stage, an inability to interbreed or at least produce fertile offspring. The frequency of hybridization, as well as the possibility for the transfer of genetic material between species, perhaps by the agency of bacteria, demonstrate that species need not be watertight entities, at least genetically. Forms transitional between species can be observed today, and can be inferred to have existed in the past. Nevertheless, the net result is very far from a seamless tapestry of form that would allow an investigator to read the Tree of Life simply by finding the intermediates—living and extinct—that in principle connect all species. On the contrary, biologists are much more impressed by the discreteness of organic form, and the general absence of intermediates.

Here, therefore, lies an important area of tension in the study of evolution. On the one hand the diversity of life can be read from an essentialist point of view, one whose vocabulary will include words such as body plan (or Bauplan) and archetype. In their more far-flung moments of comparison, proponents will take an effectively Platonic view that organic form reflects some sort of universal order, akin to the ideal solids of Platonic metaphysics. In this essentialist view the implication is that organic diversity is imposed, rather than evolved. It will also be clear that the essentialist views could be compatible with those that seek evidence for special creation in organic form. In marked contrast is an alternative viewpoint of evolutionary processes that might be linked to the famous Heraclitean flux of continuous change. In one sense this must be uncontroversial because, barring appeals to hopeful monsters testing their saltatory abilities, the facts of evolution point to building up on previously available organic designs in a gradualistic manner.

Whatever is in dispute about evolution, it is not the derivation of one type from another. But if we accept the reality of transitions do we not have to explain why large sections of potential morphospace remain unoccupied? If this is the correct analysis, as indeed it appears to be, then a more profound problem emerges as to whether such vacancies reflect lack of chance or opportunity, or whether (as seems more plausible to most Darwinian biologists) some zones of organic form (or morphospace) are effectively impossible to colonize because any organism occupying them would be seriously maladapted. In an ideal case such regions of morphospace are described mathematically. One such example is given in Chapter 8, where an example is taken from the extinct trilobites. In many other cases a precise mathematical description that defines the morphospace occupied by a group of organisms remains a very challenging prospect, but headway has been made. One of the best-known examples concerns the growth and hence geometry of the shells secreted by the molluscs, a group familiar from animals such as the garden snail and edible mussel. Although not immediately apparent, the geometry of nearly all mollusc shells can be reduced to several simple equations that together describe their various shapes.[4] What this means is that any point in mollusc shell morphospace can be defined according to a given solution of these equations. Not surprisingly, when applied to the real world such an analysis shows some regions of morphospace to be thickly populated by shell types that are relatively familiar. Other zones, however, are more or less empty. In these latter cases, the equations can be readily used to visualize the hypothetical shell shape we would find if this region of morphospace was occupied, but somehow they look 'wrong'. Such regions of morphospace housing these aberrant shells need not be entirely vacant, but the most likely explanation is that such forms are (or were) at a serious selective disadvantage for reasons such as mechanical weakness or vulnerability to predation.

To return to the specifics of organic evolution. It is generally accepted that the origins of divergence of form are coincident with the processes of speciation itself. Although it may be a mistake to think of speciation as a single process, the end results of these processes seem to be much the same. Let us accept then, if only for the sake of the argument, that not only are species discrete entities, but that they arose from pre-existing forms from which they differed in some material aspect. The central question is: are the processes of speciation in themselves sufficient to explain the pattern of life that we see today or at any time in the geological past? For nearly all biologists the fact of speciation is not in dispute, but its role in driving evolution is much more contentious. For those who do not accept speciation as the main motor of organic diversity, there are broadly two approaches. There are those who look to the molecular dynamics of the genome, as against those who seek some wider view that transcends the species. Thus, according to a number of molecular biologists the crux of the investigation needs to move to the genome and the reorganization and reshuffling of pieces of DNA, the molecular units of heredity. It is certainly realized that the genome is much more dynamic than was once thought. For example, there are large variations in the amount of DNA in different species, and it is still far from clear why some organisms have such huge excesses in DNA. There is little connection to complexity: humans for example have relatively modest amounts of DNA in each cell, but to dismiss—as some have done—the apparently excess DNA as 'junk' may be too simplistic. Not only can the amount of DNA in the chromosomes be dramatically increased, but in addition genes can be shuffled, moved around, or duplicated. There is also evidence for transfer of genetic material within the cell, notably moving DNA from the organelles known as mitochondria (which house their own separate circular chromosome) to the main storehouse of DNA in the chromosomes housed in the nucleus. In itself this activity need not be under the scrutiny of natural selection, even if the end result, the expressed phenotype, is moulded and constrained by the classical Darwinian principles of variation followed by selective culling. At the other end of the spectrum are those who argue that it is the evolutionary processes operating above the level of species that are unjustly neglected. There has been particular interest in a mechanism referred to as species selection, which in outline states that a propensity to speciation, in itself unrelated to the operation of natural selection, will favour one clade, that is, a set of species sharing a common ancestor, over another clade. While the principle of species selection appears to be logical, there are to date very few case-examples to suggest that it is of particular significance.

There are other aspects to organic evolution that are certainly not ignored, but perhaps still receive insufficient emphasis. One is the influence of the environment. This would seem to be unremarkable, until it is realized that much of current thinking seems to be firmly embedded in a uniformitarian framework, that is, it assumes that present-day conditions are a sufficient guide to understanding past worlds. In some ways this must be true: the sun shines, water is wet, and things fall out of trees. But in other ways the Earth has clearly changed dramatically. It appears that in the past 600 million years the composition of the atmosphere, notably in terms of oxygen and carbon dioxide, has changed significantly.[5] Times of elevated oxygen levels, for example, coincide with gigantism and the development of flight in some animals. There is a suspicion that there is a causal connection. Here is another example. Further back in time the Moon was probably much closer to the Earth.[6] Because of the inverse square law of gravitational attraction, the proximity of the Moon would then have generated immense tides. What effect did these have on primitive life? Could this explain, in part, the sluggishness of organic evolution at this time? This might not be the only environmental constraint. Some workers have suggested that early in the history of the Earth surface temperatures were significantly elevated,[7] and this too could have exerted a powerful brake on organic diversification.

Nor need the controls on evolution be exerted by environment alone. Barring sudden catastrophes, such as the arrival of a giant meteorite, most environmental factors will change at an imperceptible rate when compared with the generation times of living populations. But evolution proceeds not only in a real physical world, but in a biological arena. It would be simplistic to imagine that species are 'locked in' to an ecological framework, but the communities and biomes which they occupy must exert some degree of constraint.

It will be clear by now that although the Darwinian framework provides the logical underpinning to explain organic evolution, the actual pattern of life we observe may require a more complex set of explanations. Those who believe that their viewpoint is being neglected may be strident in their claims. Perhaps one reason for the continuing debate is that as a whole the various mechanisms proposed are each eminently reasonable. It is the problem of deciding if one such mechanism deserves primacy of effect, or whether the question 'Why do organisms evolve?' is unanswerable until one specifies the mechanism and the level at which it may operate. In such a large and complex field, the main strands of debate, and sometimes enquiry, are accordingly difficult to disentangle, not least because among some of the main proponents there are often broad areas of agreement. Indeed, some generate an aura of apparent accommodation by stressing their plurality of approach. On closer examination, however, this sometimes transpires to be skin-deep. Moreover, those with ideological training know that the tactics of persuasion may be assisted by the invention of key phrases that demonize the opposition.

Who then are the main proponents? Because of his earlier discussion of the Burgess Shale fauna, it is essential to review the contributions of Stephen Gould. But before doing so it is necessary to introduce briefly those who would regard their view of the evolutionary process as more or less antithetical to that of Gould. This latter group can be labelled, I think fairly, as hard Darwinists. One spokesman, Daniel Dennett, has elevated the Darwinian method to what is effectively a universal principle.[8] The acid test of such a claim is whether such a formula can explain what are presently regarded as the most fundamental and least tractable of problems, notably those of cosmology and the early history of the Universe and the onset of consciousness. For many this is taking the principle too far, and it is certainly the case that the entire philosophy is strongly materialist. In terms of organic evolution nowhere is this more evident than in the vigorous advocacy of Richard Dawkins. In a series of polemical, but carefully argued and vividly expressed books, Dawkins has unremittingly pursued the consequences—as he sees them—of the Darwinian world picture. Although set in an adaptationist landscape, a rolling and sometimes mountainous terrane that encompasses not only form and function but also behaviour, his fundamental point of reference is the primacy of the gene. In this way, Dawkins takes a highly reductionist approach. Not surprisingly, however clear the articulation, this programme has generated controversy and unease because of a sense in which the richness and diversity of evolution are being forced into an atomistic mould. Dawkins would probably reply that he is only seeking the underpinnings of the evolutionary process, upon which all else depends.

It is certainly the case that recent research into the developmental processes in animals has been little short of spectacular. At first sight these results seem to be consistent with the primacy of the gene. In a number of instances it is clear that a specific gene is associated with the expression of a complex anatomical feature. One of the best-known examples involves a so-called master-control gene which plays a key role in the formation of eyes.[9] In a classic but disturbing experiment, the application of this gene to the fruit-fly led to an ectopic expression, that is, to eyes growing on various parts of the body. But this and similar genes hold further surprises that suggest the story to be more complicated. First, it transpires that the same gene (*Pax-6* and its homologue *eyeless*) is employed not only in flies and other insects to build their characteristic compound eyes, but also in vertebrates. The eyes with which you read this page result in one sense from the activity of the same gene. Yet, despite the fact that both are light-receiving organs, there are profound differences between the eyes of fly and Man. On further reflection this need not surprise us. Most probably the *Pax-6* gene is very ancient. It almost certainly predates the animal, presumably some sort of worm, that about 600 million years ago represented the common ancestor of flies and humans. Indeed *Pax-6* may predate the earliest animals. This is because its function is to construct a light-sensitive unit, and such structures are well known in a number of the more primitive single-celled organisms whose origins almost certainly predate the animals. Equally important the recognition of *Pax-6* in arthropods (flies) and vertebrates (humans) is good evidence that they are indeed related, but it tells us nothing about the manifest differences between the eyes with which we see the fly, and the eyes of the fly which observe us as we advance with rolled newspaper in hand.

It is in this manner that Dawkins's world view is not so much wrong, as simply seriously incomplete. While few doubt that the development of form is underwritten by the genes, at the moment we have almost no idea how form actually emerges from the genetic code. In his enjoyable book *The shape of life*[10] the American evolutionary biologist Rudy Raff is bald in his assessment: 'The central problem is finding the mechanisms that connect genes and developmental processes to morphological evolution' (p. 430). One puzzling aspect, for example, is that species with very similar adult forms may reach this final stage via markedly different developmental pathways. These so-called trajectories may in themselves have adaptive significance, and no doubt different sets of genes swing into action at different times. In addition, seemingly major contrasts in anatomical arrangement may well depend on trivial genetic differences. Until, however, we learn what these are, we shall remain uninformed about the actual mechanisms whereby the shape of life is moulded. It is certainly difficult to see how the severely reductionist approach of Dawkins will continue to provide the most satisfactory strategy. Indeed, what has quite unexpectedly emerged is how seemingly very different organisms have in common fundamentally the same genetic information. Here is perhaps the central paradox of genes and evolution: vast contrasts in morphology and behaviour need have no corresponding differences in the genetic code.

Perhaps a suitable analogy to explain the short-falls of Dawkins's account of evolution is to think of an oil painting. In this analogy Dawkins has explained the nature and range of pigments; how the extraordinary azure colour was obtained, what effect cobalt has, and so on. But the description is quite unable to account for the picture itself. This view of evolution is incomplete and therefore fails in its side-stepping of how information (the genetic code) gives rise to phenotype, and by what mechanisms. Organisms are more than the sum of their parts, and we may also note in passing that the world depicted by Dawkins has lost all sense of transcendence.

In such a multifarious subject as evolution, it is certainly possible to identify camps (and outposts), but it is less easy to arrange them into a linear spectrum, let alone a simple polarity. Yet, if there is some sort of antithesis to Dawkins's portrayal of evolution, it is a yet stranger world inhabited by Stephen J. Gould, who rivals Dawkins as a popularizer of evolutionary biology. At first sight Gould's construction is much richer, especially in its appeals to a plurality of mechanisms and forces. But it is also a less constant world, or at least one where emphases and priorities shift. The world picture offered by Dawkins, as I have suggested above, is not so much wrong as simply too narrow and one-dimensional. The one presented by Gould is much more difficult to encompass, but despite its apparent vitality, I would argue that it is much more deeply flawed. Because the faunas of the Cambrian, and especially the Burgess Shale, have taken a key role in some of Gould's more recent perorations, notably in the book *Wonderful life*, it is necessary to take into account the general background of his view of life, and so its evolution.

To start with, Gould does not attempt to deny the importance of the Darwinian explanation. And indeed why should he? Some of the most cogent and readable explanations of these evolutionary principles are compelling and fascinating, especially those concerning the manner in which complex structures are 'jury-rigged' from pre-existing structures in an apparently contrived way, the nature of which clearly reveals the deep historical imprint of evolutionary activity. But Gould has also not ceased to champion the notion that the Darwinian explanation is in some way incomplete. It is hardly surprising that he has found himself at loggerheads with Dawkins. Again and again Gould has been seen to charge into battle, sometimes hardly visible in the struggling mass. Strangely immune to seemingly lethal lunges he finally re-emerges. Eventually the dust and confusion die down. Gould announces to the awestruck onlookers that our present understanding of evolutionary processes is dangerously deficient and the theory is perhaps in its death throes. We look beyond the exponent of doom, and there standing in the sunlight is the edifice of evolutionary theory, little changed. One source of unease in Gould's writing is what appears to some people as the fine line between argument and rhetoric. Thus, a favourite rallying cry of his was to label the neo-Darwinian programme, largely built on the population genetics of Morgan and Dobhansky and the mathematics of Fisher, as hopelessly sclerotic: what Gould famously labelled as 'the hardening of the synthesis'. This was a master stroke of invective, and is perhaps reminiscent of the political tactic of picking a resonant phrase to box in and demonize one's opponents. But is it a fair comment? There was only a 'hardening' inasmuch as what the neo-Darwinian school set out to do was immensely successful, and was pursued with vigour. Did it stifle research? If neo-Darwinians turned their collective back on a much-vaunted plurality of alternative evolutionary mechanisms, were they ultimately so unwise? It is significant that the recent dramatic advances in developmental biology can be directly traced to the painstaking work of these earlier neo-Darwinian geneticists. Not only that, but the repeated invitations to reinstate such individuals as Richard Goldschmidt and Otto Schindewolf from being isolated voices in the wilderness to occupy a favoured place in the pantheon of evolutionary biology have quite simply failed. That both these individuals made important contributions is not in dispute, but at the time were they ever a serious threat to our understanding of the evolutionary process? Although less commented upon than Goldschmidt, whose work on butterflies has been overshadowed by his celebrated leap into macroevolutionary thought by the agency of his much-discussed 'hopeful monsters', Schindewolf[11] is also an interesting case history. Embedded in Spenglerian cyclicity, whereby groups of organisms contained the seeds of their disaster and from high triumph descended into decadence and rottenness, his scientific influence in Germany was enormous, and baneful. A rather sinister combination of autodictat and adherence to a flawed philosophy led German palaeontology into a cul-de-sac of sterile macroevolutionary speculation and an anti-Darwinian attitude that persisted for many years after the overthrow of the Nazis.

Such is the complexity of evolutionary discussion that it would not be fair to dismiss what are now generally thought of as hopeless cases without a fair hearing. That evolution is rich in unsolved problems is not in dispute. It is certainly true that Gould's enthusiastic promulgation of various alternatives to evolutionary orthodoxies has made the guardians of neo-Darwinism look more carefully at their received truths. These alternatives have generated healthy debate. There needs, however, to come a time not only for summation and the taking of stock, but also to enquire whether other problems of evolution remain neglected. Take the case of adaptation, a key element in the Darwinian framework. That it exists is not in dispute, but is it crucial to our wider understanding of evolution? After all, if combinations of characters and traits can 'slip past' the scrutiny of natural selection, then perhaps the architecture will reveal unexpected riches. And it was by a characteristically inventive, but as we can now see flawed, metaphor that Gould started a debate on the importance of adaptation that now looks to be increasingly misplaced. He fired the first shot in his paper (with R.C. Lewontin)[12] on the spandrels of the Doges' chapel in Venice, the famous San Marco. (A spandrel can be defined in more than one way, but here it can be regarded as the roughly triangular space between the shoulders of two adjacent arches and the horizontal line immediately above their heads.) The argument that Gould and Lewontin put forward was that just as these architectural features are incidental to the design of the building, so organisms also house their own 'spandrels', which are similarly without adaptational significance. A supposed architectural by-product was taken as the introduction to a polemic on the dangers of viewing the world through adaptationist spectacles. But in fact Gould and Lewontin's analysis is fundamentally flawed. Spandrels are very far from being incidental by-products of construction and are central to design and safety.[13] It may be no accident that the almost universal human admiration of adaptation in the organic world is in some ways echoed in such buildings as San Marco. The spandrels, or more properly pendentives, house some of the glowing and mysterious Byzantine-inspired mosaics that draw the observer towards a deeper contemplation of Christian faith.[14] Moreover, is not much of our disenchantment with the barbarity of much recent architecture due to this banishment of the numinous?[15]

The case of the spandrels is one of the better known of Gould's evolutionary perspectives, and is perhaps overshadowed only by the hypothesis of punctuated equilibria.[16] Nevertheless, despite some shifts in emphasis, the underlying ideological agenda of Gould has always been fairly clear. Even where there has been a shift in thinking, it might be argued that in general the discussions were reflecting a particular world-view that at the least was sympathetic to the greatest of twentieth-century pseudo-religions, Marxism. Thus at one stage an influential group of American biologists was interested in exploring a so-called nomothetic view of evolution.[17] This was an attempt, perhaps futile, to seek general laws of evolution, which if discovered might allow the practitioners to claim that evolutionary biology was a 'hard' science, comparable in some sense to chemistry and physics. As is well known, the Marxist agenda has long sought 'laws' of history, principally linked to certain inevitable outcomes that strangely favoured those fortunate enough to have formulated the 'laws' in the first place. There is, of course, no suggestion that the hegemony of an ideology is to be transferred to the inevitability of a certain view of biology. My point rather is that the nomothetic investigation of historical sciences may reveal some interesting parallels. In any event, so far as evolutionary biology is concerned this programme has been effectively abandoned; apart, that is, from a small group of anti-Darwinians who have pursued the enterprise in the rather different direction of explaining organismal form by various underlying 'forces'.[18] In the meantime, of course, there has been a spectacular growth of interest in the operation of mathematical systems of non-linear dynamics, popularly referred to as chaos theory.

In more recent years Gould has promulgated a rather different set of notions that emphasize the role of the contingent in evolution. At first sight it is quite difficult to decide whether any of this needs to be taken seriously—until, that is, the underlying message is decoded. It is indeed somewhat surprising that the operation of contingency needs any comment at all. After all, if St Thomas Aquinas had no difficulty in reconciling the order of a Universe stemming from the Act of a Creator,[19] part of which entailed a contingent world, then we might wonder how those involved with the more mundane role of explaining evolution could sense that contingent events had been an overlooked part of the puzzle. If thorough-going theists, who traditionally have been supposed to be hostile to the scientific theories of evolution, are content to accept contingency, then one might presume that its operation in the history of life would pass unremarked. And so it might, until it was seized upon by Gould as a point worth serious discussion. In brief his argument, largely using the Burgess Shale faunas,[20] was that the range of variation in the Cambrian was so huge and the end results in terms of the diversity of today's world so restricted that the history could be regarded as one colossal lottery. Forget the big battalions, inspired leadership, the idiosyncrasy of genius, the professionalism of the academies, or any of the other factors that are routinely used to explain the twists and turns of human history, and by analogy the history of life: quite possibly they are relevant to the human condition, but no such correspondence existed in the natural world. Here chance reigned supreme, with the corollary that what to us constitutes the utterly familiar was in principle no more inevitable than a million other outcomes, ones in which humans would assuredly play no part. So much for the flights of rhetoric. Here, nevertheless, we are with one state of affairs— the world around us. How could we ever show that a plenitude of alternatives was equally likely, with the important corollary that nothing like us humans would be there in this imaginary world, either to ponder or to celebrate?

I presume that the best test of this supposition would be the discovery of a distant planet, sufficiently Earth-like to support some sort of animal life. In Chapter 9 I note that the likelihood of extraterrestrial life itself, let alone anything remotely like a human, may be much more remote than is popularly supposed. But in the immediate terms of discussing the outcomes of alternative histories, quite possibly with only marginal differences in the starting conditions, the question of whether there is or is not extraterrestrial life may not be too material to the argument. In one sense the experiment of alien life has been carried out, but here on Earth. Thus, although there is of course an evolutionary continuity in the history of life, it is also the case that not only are nearly all the species that have ever lived now extinct, but entire ecosystems have also vanished. In these past worlds there was much that was novel and has no counterpart today. But it is also true that much is familiar. This is not so much to do with evolutionary continuity, but the phenomenon known as convergence. Any textbook of evolution that fails to mention convergence would be guilty of serious dereliction. Yet despite the classic examples, which vary from the anecdotal to the closely argued, the study of convergence and the constraints of form, have, I believe, never been the subject of a single synthesis.[21] There are several reasons for this. One is its simple ubiquity: convergence is taken for granted. Another is the problem of formulating a precise metric of convergence. Famously, the marine reptiles known as ichthyosaurs are remarkably similar to the living dolphins; but are the convergences only superficial or of deep significance? Convergence is seldom precise. In addition, to identify convergence one must know the evolutionary tree that depicts both the interrelationships and ancestral conditions. But this can only be done on the basis of similarities of organization, be they anatomical, behavioural, or molecular.[22] Thereby lies the risk of becoming trapped in a circular argument: are organisms similar because they have converged or because they are descendants of a common ancestor? In terms of specifics this remains a very serious problem, but in terms of generalities the problem evaporates because no matter what evolutionary tree is chosen, convergent features almost invariably emerge. The reason for discussing convergence here is that its recognition effectively undermines the main plank of Gould's argument on the role of contingent processes in shaping the tree of life and thereby determining the outcome at any one time. Put simply, contingency is inevitable, but unremarkable. It need not provoke discussion, because it matters not. There are not an unlimited number of ways of doing something. For all its exuberance, the forms of life are restricted and channelled.

For the great majority of biologists such a conclusion will hardly be surprising. The agenda, however, once again is ideological, because the discussions on contingency versus constraint seem to be more to provide the background and focus of a very specific problem, that is the rise of human intelligence. Gould's view is unequivocal. The likelihood of Man evolving on any other planet is extraordinarily unlikely. To paraphrase: if the history of evolution were to be repeated, the world would teem with myriad forms of life (note that the contingent likelihood of the origin of life itself goes through on the nod), but certainly no humans. As stated, this seems to be entirely unremarkable, although again it presupposes that the constraints are weak. It is not, however, the point. What we are interested in is not the origin, destiny, or fate of a particular lineage, but the likelihood of the emergence of a particular property, say consciousness. Here the reality of convergence suggests that the tape of life, to use Gould's metaphor, can be run as many times as we like and in principle intelligence will surely emerge. On our planet we see it in molluscs (octopus) and mammals (Man). It might still be objected that the properties expressed in Man have a uniqueness without precise parallel. This may be a distortion of the time in which we fortuitously find ourselves; what was rare in the last four thousand million years of evolutionary history might be common in the next four thousand million years. Weak support for this argument might come from the most closely related species to us, the Neandertals. Perhaps independently they developed some sort of sense of an afterlife, at least to judge from their practice of deliberate burial. Materialists will scoff at this as a shared delusion, but there are metaphysical alternatives that are perhaps more fruitful.

But Gould's arguments on the quirkiness of human intelligence are not only presented as part of an evolutionary argument, but also I believe to buttress an ideological viewpoint. In brief, his assessment of Man as an evolutionary accident is to lead us into a libertarian attitude whereby, by virtue of a cosmic accident, we, and we alone, have no choice but to take responsibility for our own destiny and mould it to our desire. At the very least, the activities of the last century as one of unrestricted political experimentation should give us pause for thought. The implication of an evolutionary process transcending the scientific evidence does indeed provide a metaphysic, albeit one that is etiolated and impoverished, but it should be decisively rejected. We do indeed have a choice, and we can exercise our free will. We might be a product of the biosphere, but it is one with which we are charged to exercise stewardship. We might do better to accept our intelligence as a gift, and it may be a mistake to imagine that we shall not be called to account.

As I noted above, we muddy the waters of the debate if we fail to acknowledge that the processes of evolution have metaphysical implications for us. This is because uniquely there is inherent in our human situation the possibility of transcendence. The fact that we arrived here via an immensely long string of species that originated in something like *Pikaia* rather than some other crepuscular blob is a wonderful scientific story, but it is hardly material to our present condition.

Notes on Chapter 1

1. The story of the interpretation of the Devonian fossil *Acanthostega* and its key role in the understanding of how tetrapods emerged from a group of fish will be seen as one of the classic triumphs of palaeontology in this century. A detailed description of the postcranial anatomy is given by M.I. Coates (*Transactions of the Royal Society of Edinburgh: Earth Sciences*, Vol. 87, pp. 363–421 [1996]). Coates was a key collaborator with the leader of the project, Jenny Clack. This paper provides a useful bibliography, which includes two key items published in *Nature* on polydactyly (Vol. 347, pp. 66–9 [1990]) and the recognition of fish-like gills (Vol. 352, pp. 234–6 [1991]).

2. As is explained later (Chapters 3 and 4), there is compelling evidence that the Burgess Shale fauna owes its exquisite preservation in part to down-slope trans-port in turbid mudflows that led not only to its catastrophic burial but also to deposition on an area of sea floor depleted in oxygen and perhaps rich in hydro-gen sulphide. These two factors, rapid burial and a toxic environment, would have excluded scavengers and must be part of an explanation for the sensational preservation of the Burgess Shale fossils. But are they alone sufficient? Probably not, because bacteria would have continued to have flourished on decaying car-casses, eventually breaking them down completely. In an interesting paper N.J. Butterfield (*Lethaia*, Vol. 28, pp. 1–13 [1995]) has proposed that the key factor was a particular clay mineralogy that interrupted the enzymatic activity of the bacteria and thereby suspended the processes of rotting. Fossil preservation in the Burgess Shale remains a subject for controversy, and the debate has continued in *Lethaia* with comments by K.M. Towe (Vol. 29, pp. 107–8 [1996]) and a robust reply by Butterfield (Vol. 29, pp. 109–12 [1996]).

3. These problems cannot be ignored, and in a culture that has almost entirely lost its way scientists should not be so naïve as to think that today's literary aspirations will not colour their outlook. Thus, while the study of the interactions of science in the nineteenth century and the parallel creative expression of paintings, poetry, and novels are yielding rich dividends, the gulf between science and the arts in this century appears to be far deeper as each appears to pursue apparently unrelated programmes. Yet, I suspect that the creeping relativism seen in so much of the arts and philosophy, linked to the poisonous ideas of such individuals as Derrida, are far more influential than is widely acknowledged. Is it any co-incidence that some of the most celebrated of the popularizers in science, such as Stephen Hawking and Richard Dawkins, offer us such arid manifestos? Even a closer reading of the essays of the last member of this triumvirate, Stephen Gould, beneath the hyperbole and enthusiasm, reveals a much bleaker message (see my review of *Bully for Brontosaurus* in the *Times Literary Supplement*, 13 December 1991, p. 6). Recovery of the connections between the arts and science is one of our greatest, if unacknowledged, challenges. One step in this direction will be a re-examination of literary theory, and highly recommended as a vigorous antidote to present trends is the book by Raymond Tallis entitled *Not Saussure: a critique of post-Saussurean literary theory* (Macmillan, London, 1988). A further exploration of the wider topic of the place of science by the same author is *Newton's Sleep: the two cultures and the two kingdoms* (St Martin's Press, New York, 1995), although in this case I find much to disagree with.

4. The analysis of the geometric coiling of mollusc shells was first articulated more than 150 years ago by H. Moseley and slightly later by C.F. Naumann. Its study was entirely revitalized by the outstanding studies of D.M. Raup, especially with his two papers in *Journal of Paleontology* (Vol. 40, pp. 1178–90 [1966] and Vol. 41, pp. 43–65 [1967]).

5. I expressed some ideas in this area of non-uniformitarian palaeoecology in an article in *Trends in Ecology and Evolution* (Vol. 10, pp. 290–94 [1995]). At

about the same time a paper by J.B. Graham and others (*Nature*, Vol. 375, pp. 117–20 [1995]) explored in more detail the specific influences that elevated levels of atmospheric oxygen may have had on the late Palaeozoic world.

6. The argument that the Earth and Moon were once much closer, and subsequently have moved apart principally because of the transfer of angular momentum, is based on astronomical calculations. Associated with this idea is the proposal that the increasing separation of the Earth and Moon is linked to a slowing of the Earth's rate of rotation and hence to a smaller number of days in each year. An intriguing test of this hypothesis is available from fossils that not only record daily increments of growth but also have identifiable yearly intervals. An elegant analysis of some Proterozoic stromatolites (*c.* 750 Ma old) from central Australia by J.P. Vanyo and S.M. Awramik (*Precambrian Research*, Vol. 29, pp. 121–42 [1985]) provides a rather precise estimate of the number of days in the year, giving a figure in good agreement with astronomy.

7. The possibility of significantly higher surface temperatures on the early Earth is explored by D. Schwartzman and others in an article in *BioScience* (Vol. 43, pp. 390–3 [1993]). It should be pointed out that the evidence is equivocal, not least because the rock record points to significant glaciations at this time.

8. Much of Dennett's recent thinking is encapsulated in his book *Darwin's dangerous idea: evolution and the meanings of life.* (Allen Lane, London, 1995).

9. The details of the operation of the *Pax-6* and *eyeless* genes may be found in the paper by R. Quiring and others (*Science*, Vol. 265, pp. 785–9 [1994]), and G. Halder and others (*Science*, Vol. 267, pp. 1788–92 [1995]). Nor is our understanding of these genes in any way complete. Tomarev and others (*Proceedings of the National Academy of Sciences, USA*, Vol. 94, pp. 2421–26 [1997]) report the identification of the *Pax-6* gene in the squid, which being a mollusc is neither closely related to the arthropods (fly) nor chordates (mouse, human). Finding this gene in the squid is not in itself unexpected, because this animal and relatives such as the octopus have large eyes. It has long been known that these eyes are constructed in a manner remarkably similar to those of the vertebrates. What still seems remarkable, however, is that despite the differences in the molecular sequences the application of squid *Pax-6* to the fruit fly leads to the ectopic induction of eyes in the latter organism.

10. R.A. Raff. *The shape of life. Genes, development, and the evolution of animal form.* (University of Chicago Press, 1996). I address a few of the wider issues raised by Rudy Raff in his book in a review in *Trends in Genetics* (Vol. 12, pp. 430–1 [1996]).

11. One of Otto Schindewolf's key books, entitled *Basic questions in palaeontology,* was reissued in an English translation by the University of Chicago Press (1993). My review in *Trends in Ecology and Evolution* (Vol. 9, pp. 407–8 [1994]) argues that there are areas of science where ideology, then as now, takes precedence over objectivity.

12. S.J. Gould and R.C. Lewontin's paper on the metaphor of the spandrels of San Marco was published in *Proceedings of the Royal Society of London* B (Vol. 205, pp. 581–98 [1979]).

13. See R. Mark's trenchant analysis from an architectural perspective in *American Scientist* (Vol. 84, pp. 383–9 [1996]), and the subsequent biological exegesis by A.I. Houston in *Trends in Ecology and Evolution* (Vol. 12, p. 125 [1997]) where he remarks, 'I think that his [Mark's] analysis leaves the anti-adaptationist view of spandrels in ruins.' For a hard-hitting, entertaining, and intelligent review of the wider contexts of the spandrel debate, see D.C. Queller in *Quarterly Review of Biology* (Vol. 70, pp. 485–9 [1995]).

14. See O. Demus's magnificent *The mosaic decoration of San Marco Venice.* (University of Chicago Press, 1988).

15. A.T. Mann's book *Sacred architecture* (Element, Shaftesbury, Dorset, 1993) is a broadly accessible introduction, while *Morality and architecture* by D. Watkin (Clarendon Press, Oxford, 1977) is a penetrating analysis of the modern failure.

16. This latter topic is less germane to the overall aim of this book. The interested reader is well advised to consult the book *New approaches to speciation in the fossil record* (ed. D.H. Erwin and R.L. Anstey) (Columbia University Press, 1995). N. Eldredge, joint author with S.J. Gould on the original papers on punctuated equilibria, trots out the old formulae, long on theory and rhetoric, but short on actual examples. In contrast, some of the other chapters, notably that by D.H. Geary (pp. 67–86) is a model of balanced assessment that places the notions of punctuated equilibria firmly in their rightful place.

17. Key papers in this regard are those by S.J. Gould and others published in *Paleobiology* (Vol. 3, pp. 23–40 [1977]) and *Systematic Zoology* (Vol. 23, pp. 305–22 [1974]).

18. This area cannot be given justice here, but individuals influential in this field include Mae-Wan Ho and Brian Goodwin. A paper by Mae-Wan Ho in *Journal of Theoretical Biology* (Vol. 147, pp. 43–57 [1990]) provides an introduction.

19. The following translation, by Thomas Gilby (*Theological texts.* Oxford University Press, 1955, p. 97), from his Exposition on Aristotle's *Perihermenias*, makes this clear:
'All events that take place in this world, even those apparently fortuitous or casual, are comprehended in the order of divine Providence, on which fate depends. This has been foolishly denied on the assumption that the divine mind is like ours, a mistaken assumption ... The nature of the proximate causes settles whether an effect should be called necessary or contingent. Yet every effect depends on the divine will ... This cannot be said of the human will, or indeed of any created cause, for all causes, except God, are confined in a system of necessity and contingency, whether they be variable or constant in their activity. The divine will cannot fail, but we cannot therefore ascribe necessity to all its effects.'

20. The details are given in his book *Wonderful life: the Burgess Shale and the nature of history* [Norton, New York, 1989]. Many other aspects of this book by S.J. Gould are discussed in subsequent chapters.

21. By 'single synthesis' I mean a text that encompasses the range of known organisms. More specific reviews, however, are certainly available. See, for example, those by D.B. Wake in *The American Naturalist* (Vol. 138, pp. 543–67 [1991]), B.K. Hall in *Evolutionary Biology* (Vol. 29, pp. 215–61 [1996]), and J. Moore and P. Willmer (*Biological Reviews of the Cambridge Philosophical Society* (Vol. 72, pp. 1–60 [1997])).

22. Molecular convergence is widely regarded as very restricted simply because of the astronomically low odds of arranging even a short sequence of nucleotides (in DNA) or amino acids (in proteins) in the same order. This, however, is a rather special case of convergence and in other respects the evidence for such convergence is growing, with examples from a wide range that includes mutations, protein structures, nucleic acid binding molecules, transfer RNA, lysozymes, and histone genes. Key references include those by C.W. Cunningham and others (*Molecular Biology and Evolution*, Vol. 14, pp. 113–16 [1997]), S. Govindarajan and R.A. Goldstein (*Proceedings of the National Academy of Sciences, USA*, Vol. 93, pp. 3341–5 [1996]), P. Graumann and M.A. Maraherl (*BioEssays*, Vol. 18, pp. 309–15 [1996]), J.R. Macey and colleagues (*Molecular Biology and Evolution*, Vol. 14, pp. 30–9 [1997]), W. Messier and C.B. Stewart (*Nature*, Vol. 385, pp. 151–4 [1997]), K.W. Swanson and others (*Journal of Molecular Evolution*, Vol. 33, pp. 418–25 [1991]), C.B. Stewart and A.C. Wilson (*Cold Spring Harbour Symposia on Quantitative Biology*, Vol. 52, pp. 891–9 [1987]), R.S. Wells (*Proceedings of the Royal Society of London* B, Vol. 163, pp. 393–400 [1996]), J.H. Waterborg and A.J. Robertson (*Journal of Molecular Evolution*, Vol. 43, pp. 194–206 [1996]) and L. Chen and others (*Proceedings of the National Academy of Sciences, USA*, Vol 94, pp. 3817–22 [1997]; see also pp. 3485–7 for a commentary by J.M. Logsdon and W.F. Doolittle).

CHAPTER 2

Setting the scene

Overview: the structure of life

It is almost certain that within the Solar System life occurs only on Earth. In principle, both Mars and Venus may have seen the emergence of life early in their histories. But if this happened, life on these planets was doomed. It seems likely that Venus originally had oceans. These have long since boiled away, and the surface of the planet is now an inferno with a surface temperature of almost 500 °C and a lethal atmospheric cocktail of hydrochloric acid and carbon monoxide. It is just possible that life still survives on Mars. Certainly there is clear evidence for the former presence of free-running water, which in times of flood scoured out what are now dry river valleys. Recent announcements of evidence for former life in meteorites that are almost certainly Martian in origin have attracted enormous attention, yet the evidence is thin.[1] The chemical data, including measurements of the ratios of the isotopic abundances of carbon and sulphur, are difficult to reconcile. Some workers have suggested that the original temperature of the rocks was far too high for life to have existed. The supposed fossil remains, tiny vermiform objects, are even more controversial. This is not to dismiss the possibility of a Martian fossil record, but the present evidence falls far short of being compelling.[2] More importantly, if Mars ever had life it is now almost certainly extinct. All the available reports indicate that Mars is a freezing and sterile desert. Devoid of life, this planet may nevertheless yet witness the return of life in the form of colonists from Earth.

Before we consider the evolution of life on Earth we need to consider the vast stretches of geological time during which these processes and events took place. Figure 1 shows a set of pie charts to illustrate some of the principal concepts. The Solar System, formed about 4600 Ma ago, is substantially younger than the generally agreed age of the Universe. The history of the Earth is divided into four major aeons. The last of these, the Phanerozoic, is the time when fossils become visibly abundant. The Phanerozoic itself is divided into three geological eras. The Burgess Shale falls into the early Palaeozoic, with an age of about 530 Ma. The last pie chart shows how recent is the origin of the hominids in terms of geological time.

There is another way of thinking about geological time. Because light travels so slowly in relation to the size of even our galaxy, the light we see from even nearby stars takes years or centuries to reach the Earth. The light

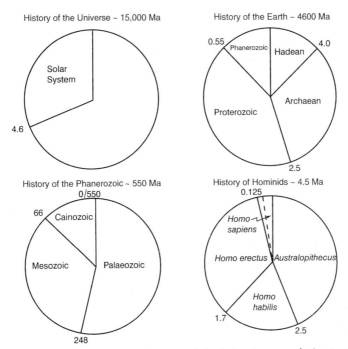

Fig. 1. Pie charts demonstrating the age of the Solar System relative to that of the Universe (upper left), the principal divisions of Earth history (upper right), the divisions of the Phanerozoic (lower left), and the time-scale of hominid evolution (lower right N.B. The upper segment refers to our species, which can be divided into an earlier 'archaic' type and (defined by the dotted line at about 125,000 years ago) anatomically modern humans).

from the nearby star of Aldebaran, for example, has taken almost 70 years to reach us. In terms of geological time the appropriate yardstick, however, has to be the distant galaxies. As Fig. 2 shows, for example, the light from a galaxy in the Hercules Cluster (numbered 2197) left about 600 Ma ago, shortly before the Burgess Shale was deposited. The light of the Centaurus Galaxy departed at about the time the great dinosaurs evolved, and shortly after light left the Virgo Galaxy the dinosaurs suffered their final and massive extinction. In contrast to these huge distances, the Andromeda Galaxy is almost on our doorstep. Its light is about 2 Ma old, and so is equivalent to the time when hominids were using their first stone tools.[3]

Exactly when and how life evolved on Earth is still uncertain, but there is no doubt that on any human timescale it is extremely ancient. A popular view is that the transition between the inanimate and the first living cells was one of imperceptible and gradual steps, that the first cells were rather simple and thereafter the history of life was one of increasing complexity. There is an element of truth in this picture, but it is really much too simplistic. Of course, we are greatly hindered by neither knowing how life evolved nor

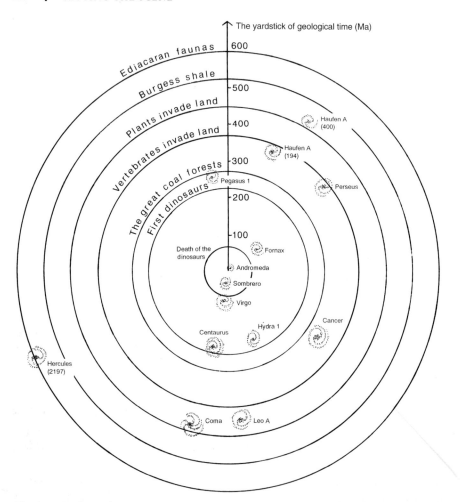

Fig. 2. Geological time and the speed of light. The circles mark important geological events from the Ediacaran faunas 600 Ma ago to the death of the dinosaurs 65 Ma ago. Also shown are a number of galaxies, the light of which has taken up to 600 Ma to reach the Earth. [Data on galactic distances and distributions largely based on M.V. Zombeck (1990). *Handbook of space astronomy and astrophysics*. Cambridge University Press.]

what the first primitive cells looked like. Perhaps the vital steps to the origin of life took place in the sediments around hydrothermal springs;[4] perhaps the first cells were most similar to still-living bacteria known as eocytes. These bacteria flourish in hot springs, inhabiting acidic water with a temperature almost at boiling. Such environments probably characterized the early Earth.

What is important to emphasize is that the complexity of life operates at several different levels. For much of the history of the Earth, marked by the intervals known as the Archaean and Proterozoic (colloquially they are referred to together as the Precambrian) (Fig. 1), life was dominated by bac-

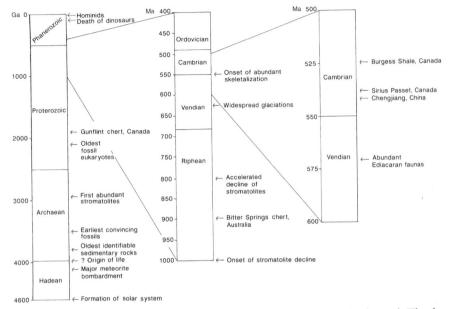

Fig. 3. Earth history with some of the principal events, mostly biological. The far right column is an expansion of the interval of geological time that encompasses the Ediacaran faunas and the Burgess Shale. Ga: billion (10^9) years; Ma: million (10^6) years.

teria. Because we tend to be impressed by large objects, bacteria seem at first sight rather unremarkable. But this is a very biased viewpoint. Bacteria may be morphologically simple, but biochemically they are exceedingly complex. They are able to attack and break down a myriad of otherwise refractory compounds. Cellulose, for example, is the basic building material of plants. Were it not for the activities of bacteria in the guts of herbivorous animals such as cows and horses, cellulose would be very difficult for them to digest. But bacteria have important limitations. For example, the morphological organization of bacteria is generally rather simple; very seldom do they form anything more complicated than chains or sheets of cells.

So how was the long march from the formation of life and the emergences of bacteria to the appearance of morphologically complex organisms, notably the animals, achieved? Only when such a background is explained can the antecedents to the Burgess Shale be understood. The context of evolution always demands a historical perspective (Fig. 3), and the route to understanding the Burgess Shale must be briefly traced from the earliest stages of the Earth's history. The Earth formed as part of the accretion of the Solar System about 4600 Ma ago. The growth of the planets and the central Sun from the rapidly rotating disc of dust and gas was violent. In particular, study of the Moon and its craters shows that early in its history it was subject to an intense bombardment by giant meteorites. There is no way in which the Earth could have avoided the same sort of bombardment; indeed it must have been much worse because the gravitational pull of the Earth is greater than that of the Moon. No direct record of this traumatic episode can be seen today on the surface of the Earth. There are indeed craters formed by giant meteorites, but all of them are substantially younger.[5] The scars of this episode of giant impacts have been erased by the processes of weathering, erosion, and the endless recycling of the crust by the processes of plate tectonics.

The early bombardment, however, had two important consequences. First, it is now believed that an important component of the infall to the surface of the Earth was cometary debris. This was rich in water and other volatiles, all essential for the presence of life. Second, some of the biggest impacts were so severe that the amount of energy released was apparently sufficient to evaporate the entire ocean and, more importantly, sterilize the Earth's surface.[6] The earliest possible evidence for life is in the form of tiny flakes of carbon, now transmuted by heat and pressure into the mineral graphite (the same substance is the basis of a pencil lead). This graphite occurs in sediments dated at about 3800 million years old.[7] At the same time the meteorite bombardment began to ease off, and it was perhaps only then that life was able to gain a foothold on the Earth.

The tiny pieces of graphite may be the earliest evidence for life, but it is not until 3500 million years ago that convincing fossils in the form of minute cells, presumably of bacteria, are recognizable (Fig. 3).[8] Remarkably it is almost three thousand million years later that the first animals appear, at least as definite fossils. What happened in this immense interval of time and why did it take evolution so long to produce animals? The outlines of the story are now becoming clear, although nobody will be surprised if there are major changes in our understanding of Precambrian life over the next few years.

All the animals, biologically known as the Metazoa, are placed in a major group of kingdom. Biologists recognize five more kingdoms. First there are the two kingdoms of the plants and the fungi. Another kingdom is known as the protistans, which comprise the seaweeds and a very wide variety of single-celled organisms, such as *Amoeba*. At first sight it is difficult to see what we animals might have in common with a cherry tree (plant), a mushroom (fungus), or an *Amoeba* (protistan). In fact it is now known that in all these kingdoms, as well as in the metazoans, there is a basic similarity in the construction of the cell. All possess what is known as the eukaryotic cell (Fig. 4). It is profoundly different from the cell type found in the two remaining kingdoms of the bacteria (the Archaebacteria and Eubacteria), which has a cellular organization referred to as prokaryotic. Eukaryotes have larger cells and a complex internal organization. This includes a nucleus (housing the DNA on the chromosomes) and specialized structures known as organelles. In addition, the plants have chloroplasts, which house the photosynthetic pigment chlorophyll. Prokaryotes, in contrast, have smaller cells

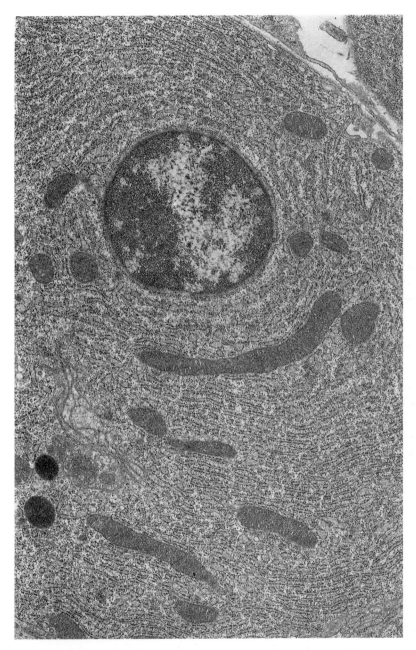

Fig. 4. Electronmicrograph of a typical eukaryote cell. Key features in this micrograph are the nucleus, circular in cross-section and containing the dark chromatin that houses the DNA, the oval to elongate mitochondria (note the internal walls known as cristae), and the labyrinth of parallel lamellae that represents the endoplasmic reticulum. The cell wall is visible in the upper right. [Photography courtesy of Jeremy Skepper, Multi-Imaging Centre, Department of Anatomy, University of Cambridge.]

(often only 0.001 mm (1 micron) across), and although they possess DNA on a chromosome it is not enclosed in a nucleus. In addition, much of the complex internal machinery, such as mitochondria and chloroplasts, is not present.

There is little doubt that all known life shares one ancestor, not least because of the shared possession of a replication system based on DNA and the presence of tiny bodies in the cell known as ribosomes (where the genetic instructions stored in the DNA code are ultimately translated into the proteins required by the cell). Nevertheless, the gulf between prokaryotes and eukaryotes is so profound that nearly all biologists take this as the basic division of life. How else can we make sense of the overwhelming diversity of life, and so bring it into some kind of order? Figure 5 shows the outline of a classification which illustrates not only the main divisions, but also the way in which we and our closest relatives (chimpanzees, gorillas) nest into the scheme. Although constructed as a set of static boxes, this diagram also encodes important evolutionary information.

Note first that the prokaryotes consist of two groups: eubacteria and the archaebacteria. There is quite substantial evidence from comparisons involving molecular biology that the eukaryotes are more closely related to the archaebacteria than to the eubacteria.[9] As mentioned above, the eukaryotes consist of four kingdoms, of which we humans belong to the animals

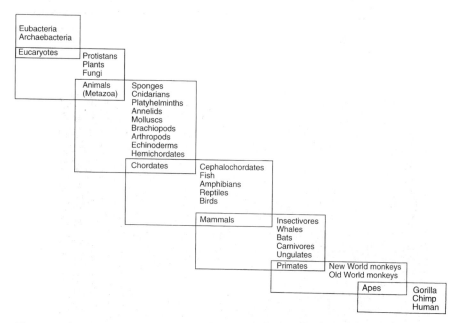

Fig. 5. The structure of life portrayed as a series of nested boxes that emphasize the position of humans. Only a few of the thousands of subdivisions are shown here. In principle, the hierarchical position of any species, living or fossil, can be established in the same manner.

(Metazoa). This kingdom in turn is subdivided into major units, known as phyla. In Fig. 5 only some of the more important ones are shown; in total there are about 35.[10] It is generally agreed that of the metazoans the sponges are the most primitive, followed by the cnidarians (which include such familiar animals as the jellyfish and corals).[11] The platyhelminthes (or flatworms) represent a major advance in organization, especially in terms of musculature, digestion, and nervous system. It is generally believed that the platyhelminthes are basal to all remaining phyla,[12] although not all of them arose directly. Those phyla known as annelids, molluscs, and brachiopods all appear to be more closely related than was generally realized in the past. In Chapter 7 the contribution of the fossil record to solving this phylogenetic problem by analysis of a strange group of Cambrian animals known as the halkieriids is reviewed. The arthropods form the most abundant component of the Burgess Shale, and remain the predominant animal group to the present day. The echinoderms can be shown, perhaps unexpectedly, to be closely related to the chordates and the rather more obscure hemichordates. Together they form a group known as the deuterostomes. The phylum of chordates is in turn subdivided into a number of classes that include the mammals. Within the mammals themselves there are further divisions, one of which is the primates. This is a quite diverse group ranging from the primitive tarsiers to apes. We humans are apes, albeit of a very peculiar type.

What we see in the Precambrian is a wide abundance of prokaryotes, mostly belonging to a group known as the cyanobacteria. These are photo-

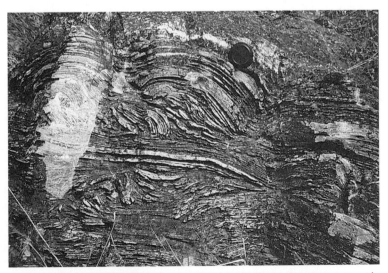

Fig. 6. A fossil stromatolite from northern China. Such structures are very abundant in Precambrian sediments, especially limestones and other carbonates.

synthetic organisms and have two important properties. First, they are the principal builders of structures known as stromatolites (Fig. 6; see also Glossary).[13] These are exceedingly abundant for most of Precambrian time, but today are generally rare. If a stromatolite is cut lengthways it usually shows a characteristic laminated structure, which is a product of the activities of the cyanobacteria. In brief, what happens is that the cyanobacteria form a dense mat, lying on the seabed and using the sunlight that penetrates the sea water for photosynthesis. Anything resting on the seabed is in danger of being buried by silts and sands, perhaps moved by currents and tides. Once a stromatolite is buried, sunlight is cut out and the mats will die. They have, however, a remarkable property of being able to glide upwards, so re-establishing themselves on the seabed. Repeated episodes of burial and upward gliding lead to the regular laminations and hence the building of a stromatolite. The other key fact about cyanobacteria is that during photosynthesis they release free oxygen. This can then accumulate in the atmosphere and oceans. With few exceptions, eukaryotes require free oxygen for respiration, and thus the evolution of eukaryotic cells was possible only by the oxygen-producing activities of the cyanobacteria.

In some ways the prokaryotic bacteria are more primitive than the eukaryotes. Nearly everyone agrees that early life was prokaryotic, and that eukaryotes evolved later. It is important to try and discover when the first eukaryotes were present, because until this more advanced type of cell appeared then there was, of course, no possibility of animals evolving. In fact the time gap between the earliest eukaryotes and the first animals appears to be enormous. What may well be primitive eukaryotic algae are known from rocks in the United States (in Michigan) that are about 2100 Ma old (Fig. 3).[14] Overall, however, rather little is known about the history of eukaryotes in the Precambrian. There is, however, clear evidence that by about 1000 million years ago there were quite advanced seaweeds.[15] Seemingly, though, animals were still absent.[16] If they were present, they would have been tiny worm-like creatures, too small to be easily fossilized.

Piecing together this story of the early evolution of life is difficult. The fossils are usually rare, their affinities may be uncertain, and the exact date of the sediments in which they are found is often contentious. The record, therefore, is sporadic and the timetable of events is subject to constant revision. Nevertheless, a coherent story is being built up, and a quite surprising level of detail is emerging, especially for the period from about 1000 million years onwards.[17] In parts of the world such as Australian, China, and Greenland the sequence of strata, deposited one above the other, provides a stratigraphic record that encompasses a considerable period of geological time; and because of the restricted degree of deformation and heating these rocks are revealing a complex biological world, albeit largely operating on a microscopic scale.

First animals: the Ediacaran conundrum

We now turn to the time when animals first appear, at least as fossils. This is between about 620 and 550 million years ago (Fig. 3). This period is usually referred to as the Vendian, after the stratigraphic sequence in Russia where rocks of this age are especially well developed. The fossils themselves, however, are often described as forming the Ediacaran assemblages,[18] taking their name from the Ediacara Hills of South Australia. Here, sediments have yielded especially rich remains of these fossils. Other localities which yield excellent examples of similar Ediacaran fossils occur in many parts of the world, including the White Sea area of northern Russia, Arctic Siberia, southeast Newfoundland, and Namibia. What bearing do these fossils have on understanding the Burgess Shale? At first sight, rather little. This is because not only is the appearance of the Ediacaran fossils rather strange (Fig. 7), but

Fig. 7. The Ediacaran fossil *Dickinsonia costata*, from South Australia. The relationships of this animal are rather uncertain, but note the prominent segments. This feature suggests that *Dickinsonia costata* may be related to groups such as the annelids.

they present an unsolved paradox in terms of their preservation. The problem is really twofold. First, in the Ediacaran organisms there is no evidence for any skeletal hard parts, similar for example to the calcareous exoskeleton of a trilobite.[19] Ediacaran fossils look as if they were effectively soft-bodied. Yet these fossils most typically occur in sediments known as sandstones and silt-stones. In normal circumstances these would be the least likely sediments in which a palaeontologist would expect or predict soft-part preservation to occur. The reason for this is rather straightforward: sandstones and siltstones tend to accumulate in areas of the sea floor that are quite turbulent and well oxygenated. Neither of these conditions is conducive to the fossilization of delicate tissues. By and large, soft-bodied fossils are found in fine-grained shales, as exemplified by the Burgess Shale.

The second aspect of the Ediacaran paradox is whether any of the fossils are actually animals. At first sight this may seem to be a ridiculous question: for many years palaeontologists have been busy comparing Ediacaran fossils to supposed modern-day equivalents, such as jellyfish or worms. A more careful scrutiny reveals, however, some significant problems. Certainly there are similarities, but they are worryingly imprecise. So what could be an alter-native? In a bold and controversial reassessment a German palaeontologist, Adolph (Dolf) Seilacher, has proposed that the resemblance between the Ediacaran fossils and any sort of animal is in fact entirely superficial.[20] He has forcibly argued that the Ediacaran biota is a separate evolutionary 'exper-iment' in body plan. Dolf Seilacher envisages the construction of the Ediacaran organisms as somewhat analogous to a mattress, with tough walls surrounding fluid-filled internal cavities. How these organisms respired and obtained their nourishment is controversial; it has even been proposed that the tissues housed symbiotic algae.

Dolf Seilacher's hypothesis is attractive because it provides a neat solution to the apparent paradox of the soft-part preservation of the fossils. He argues that it was only because of the unique construction of the body that fossiliza-tion was possible. He has coined the term Vendobionta (that is, life from the Vendian) to describe these strange fossils. The notion of the vendobionts has met with widespread approval, perhaps especially by American palaeontolo-gists. Popular ideas, however, are not always correct. I believe that this hypothesis needs to be largely scrapped. Let me explain why.

First, as is explained below (Chapter 3) there occur in the Burgess Shale remarkable fossils, looking somewhat like a fern-frond, known as *Thaumaptilon*. These are interpreted as leftovers or survivors from Ediacaran times, because very similar frond-like organisms occur in many Ediacaran assemblages (Fig. 8). The Burgess Shale fossils are important because they reveal certain details that are not readily visible in the Ediacaran specimens, which are normally preserved in rather coarse-grained sandstones. Examination of the Burgess Shale *Thaumaptilon* strongly supports its place-

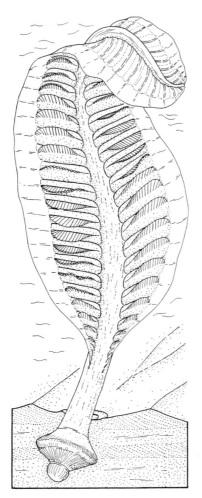

Fig. 8. The Ediacaran frond-like fossil *Charniodiscus oppositus*. The animal was embedded in the sediment, and the broad leaf housed hundreds of individual zooids that fed on particles suspended in the sea water. [Reproduced with permission from fig. 5 of R.J.F. Jenkins and J.G. Gehling (1978). A review of the frond-like fossils of the Ediacara assemblage. *Records of the South Australian Museum*, Vol. 17(23).]

ment in a group of marine animals known as the sea pens (the formal zoological name is the pennatulaceans). Sea pens are colonial animals closely related to the corals and sea anemones. Together they, and other animals such as the jellyfish and *Hydra*, are placed in phylum known as the Cnidaria. Nearly all zoologists agree that the Cnidaria are amongst the most primitive of the Metazoa. Hence, their presence in the Ediacaran assemblages is certainly to be expected.

In my opinion not only are the frond-like Ediacaran fossils cnidarians, but so too are many of the other fossils. What appears to be an intriguing absence from the Ediacaran faunas, however, are the sponges. This is surprising, because they are regarded as even more primitive, and should have appeared even earlier than the cnidarians. In fact, sponges have recently been identified both as actual fossils,[21] and also as so-called chemical fossils that have been extracted by geochemical techniques from sediments of Ediacaran age. These molecular compounds, known as biomarkers, are believed to be diagnostic of this phylum.[22] Cnidarians and sponges are both regarded as primitive, and in one sense evolutionary dead ends (see also note 10). What about the more advanced groups that must have given rise to the bulk of the Cambrian fauna? Can they be identified amongst the Ediacaran fossils? There are some candidates (Fig. 7). These fossils, which show a variety of forms, have a clear bilateral symmetry, often with a well-defined anterior end. In addition, some types may show transverse segmentation. These fossils are probably on the route leading to groups such as the arthropods and annelids.[23] Their exact place in the scheme of metazoan phylogeny is nevertheless still controversial.

This discussion of the Ediacaran faunas and the rejection of the concept of the Vendobionta is not intended to gloss over the very considerable difficulties that remain to be solved. For example, Dolf Seilacher suggests that the preservation of the Ediacaran fossils is explicable only by their possessing a unique, tough composition. This is difficult to reconcile with the frond-like fossils (and *Thaumaptilon*) being genuine sea pens, because living examples of this group, at least, do not possess an unusually thick and resistant cuticle. So why was soft-part preservation so prevalent in Ediacaran sediments? A popular suggestion is that Ediacaran communities lacked effective predators and scavengers that would otherwise consume and destroy the soft tissues. A study of the burrows and tracks (known as trace fossils) made on Ediacaran seabeds may also be relevant. Certainly there is evidence for burrowing, but the traces are relatively simple and more or less horizontal. Thus, they are effectively restricted to two dimensions. What is conspicuously absent is any indication for wide-scale churning up of the seabed by animals living in the sediment (that is, the infauna). One conclusion that may be drawn is that because burrowing was restricted it may have favoured preservation of the soft-bodied Ediacaran fossils.

Apart from the few Ediacaran survivors, such as *Thaumaptilon*, there seems to be a sharp demarcation between the strange world of Ediacaran life and the relatively familiar Cambrian fossils. The latter are typified by invertebrates such as trilobites, molluscs, brachiopods, and echinoderms. Together they exemplify the most obvious difference from Ediacaran assemblages, that is, the abrupt appearance in the Cambrian of hard skeletal parts (Fig. 3). To a palaeontologist the contrast is dramatic. In many parts of the world the sedimentary rocks outcrop in such a way as to make it easy to walk between the geological interval when Ediacaran animals flourished to a time equivalent to the Burgess Shale fauna. I could take you to eastern Siberia, southern China, south-east Newfoundland, South Australia, or many other parts of the world, but the story of organic evolution would look much the same. Rocks which took millions of years to be deposited now lie exposed in mountain ranges or along great rivers, and can be examined in a few days to reveal the outlines of the story.

In passing, it may be worth saying a little about how palaeontologists work in the field. Few areas of the world are utterly unexplored, but few regions have received anything more than cursory investigation. In many instances a new discovery will result from a regional survey by geologists, or by the enterprise of dedicated amateurs and other naturalists. Once in the area the palaeontologist almost invariably hastens to check the original report. This done, he or she will usually spend several days tramping round the area. Is the new discovery unique, or only the tip of a major new haul? What is the general setting of the fossils in terms of their sedimentology and ancient environment? Collecting itself is often hard slog; the bigger expeditions will employ teams of volunteers, and occasionally someone skilled in blasting and quarrying. The palaeontologist ought also to be sensitive to the surrounding environment, both in terms of the excavation and the rubbish of campsites. There is certainly plenty of hard work, but examining a major discovery, clutching a well-earned beer, and sitting in a region unpolluted by city lights and the continuous noise of traffic is reward enough.

Let me take you to just one such locality, the Flinders Ranges in South Australia. It is a beautiful, open country. Ranges of tall hills are separated by wide plains, in appearance not so different from the savannah of Africa, but with grasslands dotted with animals such as kangaroo and emu. The area is criss-crossed by dirt tracks, many of which lead to spectacular gorges that are ideal for tracing the geological story of the area. In these gorges are superbly exposed sequences of strata, usually tilted to an angle owing to earth movements and mountain-building. The oldest rocks in the stratigraphic sequence yield no obvious fossils, at least of animals. Slightly higher in the sequence, that is in geologically younger sediments, you enter the time of Ediacaran life. If you know where to look, Ediacaran fossils are abundant. But continue to walk along one of the steep gorges, past layer after layer of rock and so onwards through geological time. Even though the actual walk will take only an hour or two, this will be equivalent in geological terms to an interval of millions of years. If we now stop walking and examine the sediments, the Ediacaran faunas can no longer be found. They have been replaced by something much more significant: skeletons. Here is the most obvious manifestation of the Cambrian explosion.

The term 'explosion' should not be taken too literally, but in terms of evolution it is still very dramatic. What it means is the rapid diversification of animal life. 'Rapid' in this case means a few millions of years, rather than the tens or even hundreds of millions of years that are more typical when we consider evolution in the fossil record. What do we understand by the word 'Cambrian'? In essence, it is a specific period of Earth history that began about 550 Ma ago and terminated about 485 Ma ago. The Cambrian thus follows on from the last division of the Proterozoic, generally known as the Vendian, and is itself followed by the Ordovician period (Fig. 3). It takes its name from an ancient name for Wales, Cambria, where sediments of this age were first identified by the nineteenth-century geologist Adam Sedgwick. Cambrian sediments, however, are by no means restricted to Wales. There are excellent outcrops in many parts of the world, perhaps most notably in southern China, east and north Siberia, the western United States and Canada, north and east Greenland, and Australia. Taken together, examination of these sediments allows us to begin to reconstruct the outlines of the vanished Cambrian world. Its resemblance to the present day is slight. To start with, the arrangement of the continents was very different.[24] Most of them were strung out along the equator, although some land masses extended towards the South Pole. North America and Greenland thus formed an isolated continent in the tropics, straddling the equator. On the other hand, areas such as England and Wales were situated in cool southerly latitudes. Curiously, the opposite side of the world, that is the higher latitudes of the northern hemisphere, appears to have been the site of an enormous ocean, with at most only scattered islands. The fossil record confirms that the seas teemed with life, but in contrast the continents were effectively vast deserts, with at most a veneer of primitive vegetation and probably no animals. What about the climate in Cambrian times? Overall, the planet seems to have been rather warm, and it is surely significant that there is no evidence for major glaciations. This does not mean that it never snowed in the winter, but any glaciers were probably insignificant. Certainly around the tropical continents the shallow seas were often floored by limestones and other carbonates, although sands, silts, and muds also accumulated in many places. A rather remarkable feature of the Cambrian is the general rarity of volcanic deposits. Do not form the impression, however, that the Cambrian was completely unlike the world today. The seas were just as salty and the atmosphere probably had a similar composition to that of today, although perhaps with slightly lower concentrations of oxygen. As now, the processes of plate tectonics slowly moved the continents across the globe. The Earth rotated somewhat faster so that there were about 400 days in the year, but the Sun still rose every day in the east and one phase of the Moon followed the other.

Reconstructing the Cambrian world will never be easy. In many parts of the world the sediments once deposited have been worn away. Elsewhere they are deeply buried beneath younger rocks or have been altered almost beyond recognition by the metamorphic effects of heat and pressure. The arrangement of the continents has also changed very substantially. What were separate continents in the Cambrian have sometimes joined together, and all have moved to new parts of the globe. For example, from a position not that far from the South Pole in the Cambrian England has moved steadily northwards, crossing the equator about 200 Ma ago, and now lies at a latitude of 55°N.

A further problem, especially as one goes further back in geological time, is to match the histories and events that occurred on separate continents. How do we know that what happened in Wales at some point in the Middle Cambrian is more or less contemporaneous with an event in South China? In brief, for the most part we must rely on methods of geological dating that employ fossils. This is an area of geology known as biostratigraphy. It is based on the widely appreciated fact that to the first approximation sediments of the same age contain similar fossils. Problems arise because many fossils are restricted in their geographical distribution, and distant continents may house distinctive biotas. The ideal of finding a set of fossils with a global distribution is almost never realized. Thus the correlation of Cambrian sediments from one region of the world to another may be difficult.

Let us now return to the gorge in the Flinders Ranges and look more closely at the Cambrian sediments. The difference from the beds containing the Ediacaran fossils is at once apparent, because now the sediments teem with skeletal remains.[25] Many are tiny tubes; others appear to represent early molluscs. Not all these early skeletons are easy to understand. Many are patently scale-like structures, evidently once forming the skeletal coating of animals. Some belong to a group known as the halkieriids. This is a name worth remembering because we shall encounter the halkieriids again in Chapter 7.

These skeletons are the most obvious manifestation of the Cambrian explosion, but they are by no means the only evidence. The trace fossils also reveal much of interest. In comparison with those from Ediacaran sediments, those of the Cambrian are considerably more complex.[26] Not only is there a very wide variety, but many cut down through the sediment and so occupy the third dimension of the sea floor. There is a drawback, however, because the animals responsible for making the burrows are hardly ever preserved. These trace-makers were soft-bodied animals, notably the worms. What did they look like? Were they similar to those we find today in the oceans, or utterly different? Were the ecological conditions in which they lived familiar, or strange and even alien? To find out we need to turn to the heart of this book, and begin to explore the wonders of the Burgess Shale.

Notes on Chapter 2

1. The key paper is that by D.S. McKay and others (*Science*, Vol. 273, pp. 924–30 [1996]); it draws on evidence from organic compounds known as polycyclic aromatic hydrocarbons (PAHs), carbonates, crystals of magnetite, and putative

fossils. The commentary published by a British team led by M. Grady in *Nature* (Vol. 382, pp. 575–6 [1996] gives an added, albeit more circumspect, view.

2. The study of sulphur isotopes in the same Martian meteorite speaks strongly against biogenic activity being implicated in the formation of sulphides. This work is reported by C.K. Shearer and others (*Geochimica et Cosmochimica Acta*, Vol. 60, pp. 2921–6 [1996]). Work reported by D.W. Mittlefehldt (*Meteoritics*, Vol. 29, pp. 214–21 [1994]) and R.P. McSween (*Nature*, Vol. 382, pp. 49–51 [1996]) indicates that the carbonates may have formed at temperatures that were far too high for life to survive. Indications that the Martian magnetite is not biogenic, but was probably formed at temperatures in excess of 500 °C is given by J.P. Bradley and others (*Geochimica et Cosmochimica Acta*, Vol. 60, pp. 5149–55 [1996]). Not all the work, however, points to high temperatures. Later papers by J.L. Kirschvink and others (*Science*, Vol. 275, pp. 1629–33 [1997]) and by J.W. Valley and others (*Science*, Vol. 275, pp. 1633–8 [1997]) come to the opposite conclusion. Even so the debate is by no means settled and subsequent to this work E.R.D. Scott and others (*Nature*, Vol. 378, pp. 377–9 [1997]) reaffirm that the carbonates formed at elevated temperatures. The biogenicity of the polycyclic aromatic hydrocarbons has also come under renewed scrutiny (see L. Becker and others in *Geochimica et Cosmochimica Acta*, Vol. 61, pp. 475–81 [1977]). A hard-hitting review by G. Arrhenius and S. Mojzsis (*Current Biology*, Vol. 6, pp. 1213–16 [1996]) gives little support to the Martian life school, and an air of scepticism also accompanies a popular article in *New Scientist* (21/28 December 1996, p. 4) by B. Holmes.

In an interesting paper J.E. Brandenburg (*Geophysical Research Letters*, Vol. 23, pp. 961–4 [1996]) speculates that a number of organic-rich meteorites, of a type known as C1 carbonaceous, are also derived from Mars. The proposal is that early in the history of Mars, when various lines of evidence indicate that its surface was substantially wetter, there were lakes or small oceans in which organic material, derived from asteroid infall, accumulated and that later, after giant meteorite impacts, pieces were blasted off the surface of Mars to fall eventually on the Earth as carbonaceous meteorites. In this lacustrine setting it is, of course, possible that Martian life emerged in a similar way to what must have happened on the Earth about 4000 million years ago. A critique of the proposal by Brandenburg is given by A.H. Treiman (*Geophysical Research Letters*, Vol. 23, pp. 3275–6 [1996]), to which Brandenberg then replies (pp. 3377–8).

It is important to realize, however, that while the synthesis of an enormous range of simple organic compounds in unexceptional chemical conditions is relatively straightforward, their subsequent organization into replicating and membrane-bound systems may be a considerably greater hurdle and by no means an automatic consequence of surface conditions that are conducive to pre-organic synthesis.

There is one further possibility to consider. The transfer of material between planets, as has happened with Martian meteorites that have fallen to Earth, need not be a one-way process. In principle, it is conceivable that pieces of terrestrial rock have reached Mars. What if they were contaminated with bacteria? This possibility of a local panspermia is discussed by P.C.W. Davies in the book *Evolution of hydrothermal ecosystems on Earth (and Mars?)* (ed. G.R. Bock and J.A. Goode), pp. 314–17. Ciba Foundation Symposium 202 (Wiley, Chichester 1996).

3. To date the earliest evidence for stone tools is from strata 2.5 Ma. old in Ethiopia, as reported by S. Semaw and others (*Nature*, Vol. 385, pp. 333–6 [1997]).

4. The evidence for heat-tolerant or thermophilic bacteria being primitive is succinctly reviewed by E.G. Nisbet and C.M.R. Fowler in *Nature* (Vol. 382, pp. 404–5 [1996]). These authors have also explored the possible hydrothermal setting of early life, including an intriguing suggestion linking the origins of photosynthesis to the detection of infrared sources in hot springs (*Special Publications of the Geological Society of London* No. 118, pp. 239–51 [1996]).

Important information relevant to this topic can also be found in *Evolution of hydrothermal ecosystems on Earth (and Mars?)* Ciba Foundation Symposium 202 (ed. G.R. Bock and J.A. Goode), (Wiley, Chichester, 1996).

5. A helpful register of terrestrial impact structures is given by P. Hodge in his book *Meteorite craters and impact structures of the Earth* (Cambridge University Press, 1994). Relatively familiar examples include the Barringer Crater in Arizona, and the much larger Ries Crater in southern Germany. Visitors to the latter should make a point of examining the main church, St George's, at Nordlingen, a medieval town located in the crater floor. This is not only on account of its architectural merits, but also because the building stone consists of blocks of impact breccia derived from the fallout of partially molten rubble (suevite) around the crater margins. To the south-west, about 40 km away, lies the much smaller sister-crater of Steinheim. To arrive here is an extraordinary experience, because the entire crater with central uplift is immediately obvious.

   The role of extraterrestrial impacts has received enormous attention since the recognition of a major collision at Chicxulub, Mexico, 65 million years ago, which is directly implicated in the Cretaceous mass extinctions. Because there has been little success in linking any of the other mass extinctions in the Phanerozoic to impacts, the role of comets and giant meteorites colliding with the Earth has—ironically—been comparatively neglected. Calculations of the high frequency of arrival in terms of geological time and the colossal amounts of energy released by the explosion of bodies as small as 100 metres in diameter suggests that the rock record deserves fresh scrutiny in this respect.

6. Key references to this bombardment episode and its possibly controlling influence on the origin and very early evolution of life are the papers by K.A. Maher and D.J. Stevenson (*Nature*, Vol. 331, pp. 612–14 [1988]) and N.H. Sleep and others (*Nature*, Vol. 342, pp. 139–42 [1989]). In addition, the chapter (pp. 10–23) by S. Chang in the book entitled *Early life on Earth* (ed. S. Bengtson), Nobel Symposium No. 84 (Columbia University Press, New York, 1994) explains the Earth in its pre-biotic state. A number of other papers in this book provide a helpful introduction to this area.

7. A study of tiny pieces of graphite enclosed in crystals of the mineral apatite from this 3800-Ma-old deposit (at Isua, West Greenland) by S.J. Mojzsis and others (*Nature*, Vol. 384, pp. 56–59 [1996]) provides support for their organic origin on the basis of their carbon isotopes (but see the critique by J.M. Eiler and others in *Nature* (Vol. 386, p. 665 [1997]). Not all the graphite in this deposit, however, is necessarily of organic origin, and a Japanese team led by H. Naraoka has published evidence in support of a non-biogenic origin (*Chemical Geology*, Vol. 133, pp. 251–60 [1996]).

8. An update for the evidence of 3500 million-year-old life is given by J.W. Schopf (pp. 25–39) in the book *The Proterozoic biosphere: a multidisciplinary study* (ed. J.W. Schopf and C. Klein) (Cambridge University Press, 1992).

9. Key papers in this regard are those by M.C. Rivera and J.A. Lake (*Science*, Vol. 257, pp. 74–6 [1992]), P.J. Keeling and W.F. Doolittle (*Proceedings of the National Academy of Sciences, USA*, Vol. 92, pp. 5761–4 [1995]), and S.L. Baldauf and others (*Proceedings of the National Academy of Sciences, USA*, Vol. 93, pp. 7749–54 [1996]).

10. An important new venture in providing an overview of the metazoan phyla is being provided by P. Ax, of which the first volume, *Multicellular animals: a new approach to the phylogenetic order in nature* (Springer Verlag, Heidelberg 1996) is an encouraging beginning. Also very useful is the book by C. Nielsen entitled *Animal evolution: interrelationships of the living phyla* (Oxford University Press, 1995).

11. Sponges and cnidarians are so different that it is difficult to imagine how the transitions either between them, or between the cnidarians and the platyhelminthes

CHAPTER 3

The discovery of the Burgess Shale

Tantalizing hints

It was inevitable that the Burgess Shale was going to be noticed sooner or later and its fossil treasures revealed to science. The actual story of its discovery is, nevertheless, a dramatic episode in the history of palaeontology. It was an event that required both opportunity and the hand of chance. First, let us consider how that opportunity arose. If one drives across western Canada, then, having passed through the prosperous city of Calgary, it is only a few hours before one is greeted by the Rocky Mountains, literally rearing up from the rolling prairies. In these days of superhighways and long-range aircraft it is very difficult to realize that little more than a hundred years ago these mountain ranges formed an effectively impassable barrier.

As Canada established its identity as a nation, it was realized that a single country could be forged only if there were easy communications. Before the advent of cars and aircraft the connecting link had to be a railway, joining Vancouver on the west coast to the rest of Canada. Building this railway across the Canadian Rockies represents one of the great feats of engineering.[1] With the railway open, it was inevitable that nearby towns developed or grew. Some, such as Banff and Lake Louise, attracted tourists, drawn to the spectacular mountain scenery. Others, especially the small settlement at Field, were largely concerned with the running of the railway.

The first hint of the palaeontological riches of the area around Field came with the building of this railway.[2] Field lies in the shadow of Mount Stephen (Fig. 9). Fossils had been found on this mountain at least as early as 1884, by a surveyor employed by the railway. A couple of years later a geologist working for the Geological Survey of Canada, having been told about previous discoveries of fossils, climbed Mount Stephen. High above Field he discovered a vast natural storehouse of ancient life. Scattered over part of the mountain, just above the tree-line, were millions of yellowish slabs (Fig. 10), on which occurred a profusion of trilobites. These beds are known as the *Ogygopsis* shale, the name being taken from a particularly abundant trilobite. Now Cambrian rocks often yield trilobites, if not in the abundance of the *Ogygopsis* shale. But this locality also yielded many specimens of a much stranger fossil, which was to prove a harbinger for the even more extraordinary fossils from the Burgess Shale. In an early scientific description (1892) by the palaeontologist J.F. Whiteaves[3] this odd fossil was appropriately

Fig. 9.   Mount Stephen, British Columbia. The highest peak is almost 3200 metres.

Fig. 10.   The fossil beds of the *Ogygopsis* Shale on Mount Stephen. The standing figure is Dr Jim Aitken, one of the leaders of the Geological Survey of Canada Burgess Shale excavations in 1966 and 1967. The mountain in the centre of the skyline is Wapta Mountain; the one to the right and partly obscured by a tree is Mount Field. The Burgess Shale lies between these two mountains.

Fig. 11. The fossil *Anomalocaris canadensis* from the *Ogygopsis* Shale, Mount Stephen.

named *Anomalocaris*, literally the 'strange crab'. Whiteaves considered the fossil (Fig. 11) to represent the abdomen of a crustacean, which is familiar as the fleshy 'tail' of lobsters and shrimps. We will discover later in this book that this interpretation of *Anomalocaris* was wildly wrong, but it was by no means the last mistake to be made in the understanding of this strange and enigmatic fossil.

Palaeontologists were therefore beginning to become acquainted with the fossil riches of this area, but the story takes a crucial turn with the arrival of a truly remarkable man, Charles Doolittle Walcott (Fig. 12). He was one of the most able and energetic of American geologists, and in his time he was widely recognized as the leading authority on Cambrian life. Walcott had for many years been exploring the Cambrian rocks of western America and he must therefore have been well aware of the *Ogygopsis* Shale. As early as 1888 (Fig. 13) he had published a paper[4] describing some of the fossils, but these had been sent to him by collectors. In 1907 he visited the *Ogygopsis* Shale, collecting trilobites and other fossils from what seemed to be an inexhaustible resource.[5] Working on the outcrop of the *Ogygopsis* Shale it would have to be a very dedicated palaeontologist whose mind was only on the fossils: the scenery visible from Mount Stephen is truly spectacular. Perhaps as Walcott looked northwards, across the Kicking Horse Valley through

Fig. 12. Charles Doolittle Walcott, discoverer of the Burgess Shale, standing in the Walcott Quarry. [Photograph courtesy of the Smithsonian Institution (Dr D. Erwin).]

which snaked the newly completed railway, he remembered what the English geologist Henry Woodward had written in 1902.[6] Woodward had described a number of fossils from the *Ogygopsis* Shale, but in this paper he had also written, 'Mount Field, which faces Mount Stephen, remains still unexplored, but is part of the same massif, and will no doubt yield the same Cambrian fossils'. Walcott was an ambitious and curious man. Perhaps the combination of the richly fossiliferous *Ogygopsis* Shale and unexplored territory to the

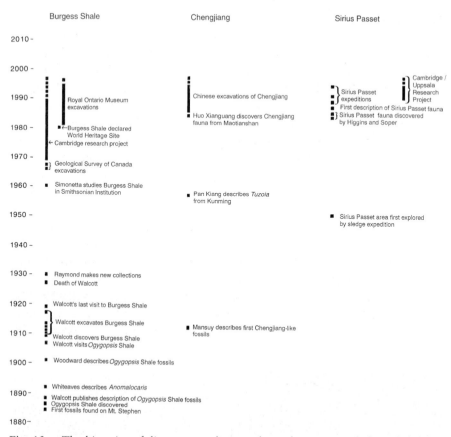

Fig. 13. The histories of discovery and research on the Burgess Shale and the two other most important soft-bodied faunas of Cambrian age: Chengjiang from South China and Sirius Passet from North Greenland.

north provided the necessary catalyst? In any event, within a year Walcott had discovered the Burgess Shale.

Walcott's triumph

It will probably never be known exactly how Walcott discovered the Burgess Shale. The old story goes as follows. Walcott's party was returning from fieldwork, travelling along the trail that runs parallel to and to the west of the enormous ridge that connects Mount Field and Wapta Mountain. As was usual at that time, everyone was on horseback. Mrs Walcott's horse stumbled on a block of rock, her husband dismounted, split open the boulder, and to everybody's amazement the blow of Walcott's hammer revealed a profusion of soft-bodied fossils. The story, however, appears to be no more than a legend, although who can be surprised that the story of the discovery of the most wonderful fossil deposit in the world was embroidered and retold? In fact Walcott's diaries show that the first of the Burgess Shale fossils were discovered by him, with his wife Helena and son Stuart, on Tuesday, 31 August 1909, or perhaps by Walcott the previous day.[7] The next few days were spent busily collecting more fossils; already the riches of the Burgess Shale were self-evident. The first fossils must have been collected from the thousands of loose pieces of shale scattered over the hillside. It was obviously a priority to discover the intact stratum, the mother-lode, from which these fossils were derived. The source of the fossils may have been found before the end of the 1909 season, which was typically curtailed by the first heavy snowfalls in early September. Walcott returned in 1910 to start major excavations (Fig. 14), which then continued until 1913, with a final season in 1917 (Fig. 13). At the end of each collecting season the fossils were taken down the trail by horse or mule to the railway at Field and thence sent to the United States. Their final destination was Washington, D.C., and specifically the Smithsonian Institution, whose director was Walcott himself.

Walcott must have been enormously excited by his discoveries. By 1912 he had published a whole series of scientific papers,[8] setting out his interpretations of this fantastic fauna, which far exceeded in quality that from the

Fig. 14. Charles Walcott excavating at the Burgess Shale quarry. [Photograph courtesy of the Smithsonian Institution (Dr D. Erwin).]

nearby *Ogygopsis* Shale. But by his last field season at the Burgess Shale Walcott was 66 years old. He had suffered the grief of his son Stuart dying in the First World War, shot down in aerial combat over France. Earlier, in 1911, his wife and loyal field companion had perished in a railway accident. Walcott himself was becoming increasingly absorbed in administration, taking a leading role in science policy and the encouragement of new research. Time for studying the Burgess Shale was slipping away. A posthumous paper published in 1931, four years after his death, is little more than a series of notes.[9] Despite his increasing commitments Walcott clearly planned to pursue his research into the Burgess Shale. Several times during my own investigations of the fossils in the Smithsonian Institution I came across his photographs alongside specimens that turned out to be completely new to science.

By 1918 Walcott was able to write that his excavation 'practically exhausts a quarry which has given the finest and largest series of Middle Cambrian fossils yet discovered.'[10] Perhaps he genuinely believed the quarry had no further potential, but I wonder if his comment was as much a smoke screen to discourage other investigators. In any event, in the world of palaeontology, the Burgess Shale was well known. By the standards of his day Walcott's publications were no disgrace. Probably to any of his contemporaries his scientific papers must have seemed to be perfectly adequate.

Even before Walcott's death, however, other Americans were taking an interest in the Burgess Shale. In particular, a Harvard professor of geology, Percy E. Raymond, ran summer field camps that included, at least in 1925 and 1927, visits to the Burgess Shale. Raymond was one of America's leading specialists on trilobites. He too had an interest in exceptionally well-preserved fossils. Like Walcott he had studied the fossilized remains of the soft tissues of trilobites, which are very seldom preserved. Much of Raymond's work was on fossils younger than the Burgess Shale, but he had also visited the *Ogygopsis* Shale and published an account of several new fossils.[11] In 1930 (Fig. 13), perhaps judging that a suitable interval had elapsed since Walcott's death, Raymond undertook a serious excavation of what has come to be known as the Walcott Quarry.[12] More importantly, perhaps, he explored the hillside and opened a new quarry about 20 m higher. The general quality of fossil preservation in this excavation, now known as the Raymond Quarry, is not as spectacular as in the lower Walcott Quarry. These discoveries were, however, important because although Walcott had discovered fossils from these higher levels, Raymond demonstrated the possibility of such exceptional preservation being more widespread than had hitherto been thought. As we shall see in Chapter 5 it is now clear that fossil preservation typical of the Burgess Shale is much more widespread than either Walcott or Raymond could possibly have imagined.

Indeed, although the really special accumulations of Burgess Shale-like fossils are uncommon, this type of exceptional preservation is really quite widespread in the Lower and Middle Cambrian. Why it should be so is not very clear, although the Canadian palaeontologist Nick Butterfield has proposed that special properties of the sediments, notably the clays, may have acted to inhibit bacterial decay.[13] Examples of special or exceptional preservation of fossils are often referred to as *Lagerstätten*, a German term that originally referred to rich mineral deposits. Fossil Lagerstätten occur in many other geological horizons younger than the Cambrian. Among the most famous are the Devonian Hunsrück Shale (Germany, *c.* 395 Ma),[14] the Mazon Creek nodules (Illinois, *c.* 295 Ma),[15] the Solnhofen Limestone (Germany, *c.* 150 Ma),[16] and the Messel Oil Shale (Germany, *c.* 50 Ma).[17]

The Cambridge campaign

The collections of Burgess Shale fossils that Raymond made were quite extensive. It is really very surprising that despite resting in the prestigious Museum of Comparative Zoology (MCZ) in Harvard for many years the fossils were almost entirely neglected. This was not quite true of Walcott's collections in the Smithsonian, because in 1960 an Italian ornithologist and ecologist, Alberto Simonetta, studied a wide range of the fossils in this museum. Both these collections are, of course, in the United States. It was in the 1960s that the Canadians, notably Digby McLaren, who was director of the Geological Survey of Canada (GSC), realized that their country might well be host to a superb fossil locality, but ironically it had almost no specimens of its own.

The first hurdle was to obtain permission to excavate the Walcott and Raymond quarries from the authorities of what was now Yoho National Park. It was also necessary to find leaders for the new project. Dr James D. (Jim) Aitken (Fig. 10) of the GSC was placed in charge of the overall programme, and it is clear that his enthusiasm and invigorating style were the key to the successful excavations. In addition, the leadership of the central purpose of the reinvestigation, the biology and ecology of the Burgess Shale animals, was assigned to Harry Blackmore Whittington. He is an Englishman, but had been long resident in the United States as a professor of palaeontology in Harvard. The GSC obtained permission for three consecutive years of scientific excavation (Fig. 15), but for various reasons the first season (in 1965) was postponed. By 1966 Harry Whittington knew he would be returning to England, to take up one of its most prestigious professorships, the Woodwardian Chair at Cambridge (which takes its name from the founder of the professorship, John Woodward, who in 1728 bequeathed his museum to the University).

Fig. 15. The Geological Survey of Canada excavations at the Burgess Shale. [Photograph courtesy of H.B. Whittington (University of Cambridge).]

Excavations continued in 1966 and 1967, but attempts to persuade the Park authorities that quarrying should be allowed in 1968 as part of the three-year agreement were frustrated. Nevertheless, a substantial collection of fossils had been made, and it was soon on its way to Cambridge. Why had Harry Whittington been chosen to lead the investigation into the Burgess Shale fossils? At the time he was invited to lead the project he was already recognized as the world's leading specialist in trilobites. Walcott's work had shown that the Burgess Shale was rich in arthropods, the phylum to which trilobites and other animals such as shrimps and flies belong. At first sight the fact that the extinct trilobites, so named because the body is divided into three lobes (one central and two flanking), are related to the more familiar arthropods such as crabs and spiders may seem surprising. There are, however, similarities that most zoologists regard as fundamental, most notably the presence of a jointed exoskeleton. Thus, as Whittington was already famous for his meticulous work on trilobites he was the logical choice to spearhead the campaign. This explains also why two other

Englishmen were also involved almost immediately in the study of the Burgess Shale. One was David Bruton, who had assisted with the excavations in 1967. The other was Christopher Hughes, newly appointed to a research post in Cambridge. Both men were specialists in trilobites, and together with Harry Whittington they devised the first assault on the Burgess Shale fauna. Whittington undertook the analysis of a small arthropod, already named by Walcott as *Marrella*.[18] There is a nice irony in this because Walcott chose the name to honour John Marr, a predecessor of Harry Whittington as a Woodwardian professor. In the meantime, David Bruton embarked on the study of a much larger arthropod *Sidneyia*,[19] named by Walcott in honour of his son Sidney. Chris Hughes took the first steps in the analysis of yet another arthropod, the eponymous *Burgessia*.[20]

When scientists name animals they almost always follow a system that was codified in the late eighteenth century by the Swedish botanist, Carl Linné. This system is known as binomial nomenclature, because two names are employed. Thus the name for our species is *Homo sapiens*. *Homo* is the generic name and *sapiens* the specific name. So the full names of the three Burgess Shale arthropods mentioned above are *Marrella splendens*, *Sidneyia inexpectans*, and *Burgessia bella*. How clearly each of the species names reveals Walcott's intense excitement: *Marrella* the splendid, *Sidneyia* the unexpected, *Burgessia* the beautiful.

The first public landmark in the reappraisal of the Burgess Shale was in Chicago in August 1969. Here, at a convention of palaeontologists, Harry Whittington gave a preliminary account of *Marrella* (Fig. 14).[21] In doing so he revealed the first fruits of the research enterprise by stating three key facts. First, the specimens of *Marrella* were exquisitely preserved, a point that could be appreciated only incompletely from Walcott's publications. (In his time the standard of photography was less impressive than today, and Walcott routinely retouched the photographs, the better to do justice to the amazing fossils.) Nevertheless, the preservation of *Marrella* showed some peculiarities. Most notable was the almost ubiquitous presence of a dark patch or stain, especially towards the posterior and sometimes the anterior end of the specimen (Fig. 16). Harry Whittington originally interpreted this stain as the remains of the innards that had been squeezed out as the overlying mud was compacted into shale. Subsequently, however, it was realized that this stain was formed by body contents oozing out into the surrounding sediment. The presence of a dark stain indicated that the dead animal had begun to rot and decay. Then, mysteriously, one or more factors had prevented decay from going any further, so allowing the delicate remains to fossilize. This, incidentally, remains one of the major unsolved mysteries of the Burgess Shale, and we still have little idea as to why the fossils are so exquisitely preserved.[22] Second, Harry Whittington not only confirmed that *Marrella* was an arthropod, but he also showed that although it had certain similarities to groups

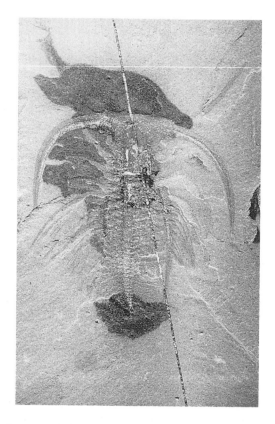

Fig. 16. The Burgess Shale arthropod *Marrella splendens*. The line crossing the specimen is a vein of the mineral chlorite.

such as trilobites and crustaceans (the former are extinct, the latter are represented today by animals such as shrimps, crabs, and lobsters), it could be placed in neither of these two groups, nor in any other major group of arthropod. In this sense *Marrella* was a zoological enigma. Third, during his study of the thousands of specimens of *Marrella* he noticed that the fossils were buried in all sorts of different orientations. In addition, he observed that the various parts of the animal, notably the limbs, were separated from each other by thin layers of sediment. Harry Whittington could only conclude that the specimens of *Marrella* had experienced turbulent transport. Carried along in mud-clouds they had been whirled around, with sediment seeping between the limbs. As the strength of the mudflow ebbed and it slowed down, so the animals were quickly deposited, together with the surrounding wet mud. This process was so rapid that the animals did not have time to adopt a stable orientation. Instead they came to rest at a variety of angles, some specimens even being buried head-first. As more and more sediment was deposited on top of the animals, so the wet muds containing *Marrella* and the other

Burgess Shale fossils were increasingly squeezed. The water was expelled, the fossils crushed, and the mud slowly turned into a tough shale.

At an early stage of these investigations, Harry Whittington made a survey of Walcott's huge collection in the Smithsonian Institution. Combined with the smaller, but still significant, collections made by the Canadians, it was clear that a comprehensive reappraisal of the Burgess Shale fauna would need a team of specialists. The initial trio of Whittington, Bruton, and Hughes would not be enough. It was decided to find at least one graduate student. That is how I became involved.[23] So far as I can remember, my first encounter with the Burgess Shale was in the Geology Department of the University of Bristol, where I was an undergraduate. Although the great bulk of Walcott's collection had remained in Washington, D.C., Walcott or his assistants had arranged for small collections of Burgess Shale fossils to be sent to many museums. In Bristol one of our teachers in palaeontology had looked out the few drawers of specimens for a laboratory practical. The fossils were not that spectacular, but my interest was sparked. As I recall I went to the geology library, looked up some of Walcott's papers, and my interest grew. I learnt that Harry Whittington was masterminding the Burgess Shale project. I wrote to him, and he invited me to come for an interview. In Cambridge the Department of Geology, as it was then known, forms part of a complex of science buildings close to the centre of the town. To get there I had first to cross the River Cam, then on through the market-place, and so to the Department. Whittington's room was magnificent, situated at the far end of the Sedgwick Museum, which occupied the entire first floor of the building. The interview went well, but he explained that another student also wanted to work on the Burgess Shale. His name was Derek Briggs. Not only had the Cambridge college to which Harry Whittington belongs, Sidney Sussex, offered a research studentship to Derek Briggs, but not unreasonably he had been given first pick of potential research topics for his Ph.D., and had opted to join the main campaign tackling the arthropods. With considerable diffidence Harry Whittington suggested that the Burgess Shale worms might be a suitable subject. That sounded fine to me, although it was only in August 1972, a month before I was due to start in Cambridge, that final confirmation of my funding was given.

I should add that even before meeting Harry Whittington for the first time I had a strong intuition that the Burgess Shale project would be exceptionally promising. How on earth can a scientist decide in advance that a project is pregnant with possibilities rather than turning into a thoroughly mundane piece of work or even a complete failure? Type of education, general background, and chance factors must all play their part, but I certainly rely on intuition and hunches. Now this could be regarded as rather embarrassing. Scientists are usually depicted as individuals rather detached from the mainstream of life, calculating, absorbed in logical thought, impatient with emotion, dedicating their lives to the precise delineation and subsequent solution of problems. In my experience this describes how a scientist justifies arriving at a successful conclusion, but does not have very much to do with the actual process of investigation. The route to a successful piece of work seems to be much more circuitous and irregular. Perhaps I am an exception, but I believe these comments are of some relevance to explaining science to non-scientists. I am not denying, of course, that science is a logical subject. One must take the greatest care in explaining how one arrives at a conclusion, and one must always think of ways of testing one's hypotheses. Nevertheless, I have little time for books with titles, imaginary in this case, such as 'How to become a world-famous scientist'. They try to describe the strategy for success, but make no mention of intuition, hunches, even sometimes dreams.

So it was now October 1972. Derek Briggs and I were all ready to embark on our respective projects: he to study an arthropod known as *Canadaspsis perfecta* (literally, 'the perfect Canadian crab'), while I was due to start on the worms. No sooner had I begun my work than I developed appendicitis, followed by a bout of septicaemia. By March of the next year, however, we had learnt enough from the material already in Cambridge, collected by the GSC expeditions. Now we were ready to travel to the Smithsonian Institution to start the trawl through Walcott's mammoth collections. By that time we had both learnt three important research techniques: photography, the preparation of the fossils, and how to make detailed drawings. Palaeontology is a very visual science, and most publications include numerous photographs of the specimens and quite often reconstructions of what the author believes the organisms looked like when alive. Whittington had discovered that photographing the Burgess Shale fossils in ultraviolet (UV) radiation, which when it comes from the Sun is of course responsible for sunburn, often produced excellent results. Many of the Burgess Shale fossils are preserved as reflective films on the surface of the shale. By tilting the fossil to the correct angle relative to an ultraviolet lamp, it was possible to take striking photographs. Working in the photographic room was an eerie experience, bathed in the purple glow of a UV lamp, and squinting through the camera to try and obtain the best results.

What about the preparation of the fossils? Often part of specimen is obscured by an area of shale that needs to be chipped away. In addition, because of the seepage of sediment between different parts of the body, it is sometimes possible to remove an overlying portion and so reveal a hitherto concealed part of the fossil's anatomy. Many of the Burgess Shale fossils are small, and often only a few square millimetres need to be excavated. To achieve what is in effect a sort of dissection, Harry Whittington devised a specially modified dental drill with a percussive action. By judicious handling, small areas of a fossil could be exposed. Sometimes this meant the destruction of pre-existing parts of the fossil. If necessary, photographs would be taken to record each stage of the drilling. Preparation of the specimens was always done using a microscope, and was a painstaking process. But it could be exciting: as the covering layer of shale was chipped away some completely unexpected feature might be slowly revealed.

Harry Whittington introduced us to another technique that might seem at first rather pointless, but was essential for the success of the project. This was the preparation of detailed drawings of the specimens (Fig. 17). There are two reasons why such drawings were necessary. First, however good the photographs may be, they cannot show all the details of a specimen. By placing drawing and photograph side by side the interpretation of the fossils is much easier to understand, particularly by the non-specialist. Second, by making a drawing one is compelled to scrutinize all areas of the specimen: one is forced

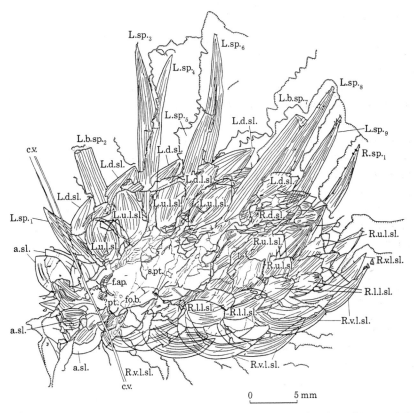

Fig. 17. A typical camera-lucida drawing of a Burgess Shale fossil. In this case it is of a well-preserved specimen of the worm *Wiwaxia corrugata* (see Fig. 43). [Reproduced with permission from fig. 105 in S. Conway Morris (1985). The Middle Cambrian metazoan *Wiwaxia corrugata* (Matthew) from the Burgess Shale and *Ogygopsis* Shale, British Columbia, Canada. *Philosophical Transactions of the Royal Society of London* B, Vol. 307, pp. 507–86.]

to make interpretations. The actual technique of drawing is very simple. The microscope has a special adaptor that throws the image of the fossil on to a sheet of paper, where it can be traced. Nevertheless, the process of preparing a drawing could be very laborious; a single specimen might take several days to draw. Rotating the specimen, placing it at different angles, and altering the intensity of the lamp were all necessary to produce a satisfactory drawing.

The time taken to study thoroughly even a single specimen may, therefore, be many days. Some species are known only from a single specimen, but others are much more abundant. *Marrella*, for example, is represented by at least 15,000 individuals.[24] Obviously not all can be studied in equal detail. But even if a specimen is subject to intense scrutiny, this is no guarantee that the interpretation is necessarily correct. It is important to understand that mistakes are inevitable in science. Naturally, one does one's best to avoid them, but if they are detected the utmost honesty is required. One should also remember that so far as science is concerned, practically everything is hypothesis. It is capable at any time of refutation, and no matter how much evidence appears to support the hypothesis it remains impossible to prove once and for all that it is true. Even things we take completely for granted are still hypotheses. All the evidence we have indicates that the Earth orbits the Sun. However, although the description of the orbit is explained very well by Newtonian mechanics, it is generally agreed that the orbit can be explained even more accurately by the theory of general relativity. But it is also agreed that in certain important respects relativity theory is still incomplete. Similarly who can doubt the existence of atoms? But in fact the story of the search for building-blocks of elementary particles that make up atoms is by no means finished. Moreover, at the atomic scale the world is organized according to quantum physics, which in many ways is utterly different from the macroscopic world with which we are familiar. How the transition from quantum to macroscopic world occurs is frankly speculative. If the fundamentals of physics are controversial, so too are many areas of biology. But we have no choice. Some scientists tread the careful path, sticking to safe subjects and performing worthy work on areas that are believed to be non-controversial. Others are more adventurous. They know that their journey through science is risky, and more often than not they will be proved wrong. But they have realized that science is a human endeavour, one that involves play and intellect. At its best it is a heady mixture.

Two mistakes

So the boxes of specimens from the Smithsonian Institution are arriving in Cambridge at regular intervals, and stacked high in the research rooms are the fruits of the excavations made a few years earlier by the Geological

Survey of Canada's collecting teams. Shelves are piled with books and journals, and the research files are also steadily accumulating: scientific reprints, negatives and photographs, camera-lucida drawings, and notes ranging from jottings to detailed descriptions of particular fossils. A typical research environment, and with the aim of producing not only a doctoral thesis, but as importantly a set of publications that will be the essential first step to any sort of academic career. But the collections are huge, and not everything can be scrutinized with equal care. There are further complications. Some species are known from only a handful of specimens, sometimes only a single fossil. In addition, the preservation of these rare specimens may be far from satisfactory. There are also other pitfalls. Scientific preconceptions dog every step of the investigation. All researchers try to be neutral, to consider the alternatives, but the basic business of science is to build hypotheses. No scheme is ever complete, certain observations remain naggingly out of context, and sometimes the crucial piece of evidence simply is not available. The hypothesis is worked upon, confidence grows, and things seem to fall into place. But is it really correct?

Walcott's main area of research into the Burgess Shale was the arthropods. This is hardly surprising, given that one of his principal interests remained the trilobites. He nevertheless described a significant proportion of the remaining fauna, including many of the worms, which were my assigned research topic. One of these worms Walcott named *Canadia sparsa*. The generic name of the animal refers simply to Canada, where the Burgess Shale is located. The specific name *sparsa* needs, however, a little more explanation. There is another Burgess Shale worm which Walcott called *Canadia spinosa*. As we shall see, it is a very interesting worm, belonging to a group known as the polychaete annelids. Representatives of this group of worms characteristically bear bundles of bristles, known as the chaetae. The exact arrangement of the bristles varies between species, but in *Canadia spinosa* the body bears prominent bundles of these chaetae; hence the specific name refers to its spinose appearance. In contrast, *Canadia sparsa* has only a few large spines; so Walcott's choice of the specific name is a direct reference to their sparseness.

Because I had chosen to study worms, it was necessary to include *Canadia sparsa*. Looking at this worm in the Smithsonian Institution, it was immediately obvious that it was not at all similar to *Canadia spinosa*. This opinion was reinforced as I excavated away some of the sediment that covered various parts of the best-preserved specimen (Fig. 18). I showed a preliminary drawing of this specimen, depicting an elongate body with seven pairs of enormous, sharp spines arising on one side of it and on the opposite side a series of flexible tentacles, to an English colleague who was also working at the time in the Smithsonian. He burst out laughing. It was clear that this animal was very different from *Canadia spinosa* and it required a new name.

Fig. 18.   The best-preserved specimen of *Hallucigenia sparsa* from the Burgess Shale.

I discussed the problem with a friend, Ken McNamara, who was also working on trilobites with Harry Whittington. We tried out various new names, and chose *Hallucigenia*[25] as a tribute to its dream-like appearance. When reconstructed this animal was depicted as being propped up on the sea floor, its seven pairs of spines embedded in the sediment and the tentacles projecting upwards, perhaps in search of food (Fig. 19a). *Hallucigenia* was such an extraordinary fossil that I began to wonder if it might be only part of a much larger animal. But this did not seem very likely because in Raymond's collection in the MCZ in Harvard there is a very strange specimen with about eighteen *Hallucigenia*s associated with a large organic mass. This strongly suggested that the *Hallucigenia* animal was a scavenger, marching across the seabed in search of decaying corpses.

The apparent weirdness of *Hallucigenia* began to be taken as the exemplar of the Burgess Shale. Life in this community came to be regarded as a Cambrian bestiary with an organic exuberance that completely overshadowed the dull ordinariness of typical Cambrian faunas. Other interpretations were, of course, suggested. In a student textbook Richard Cowen[26] speculated that *Hallucigenia* did not walk over the sea floor, but shuffled through the sediment. He wondered if *Hallucigenia* used its tentacles for locomotion and protected itself with the long spines projecting upwards out of the mud. In fact Richard Cowen was not so far wrong. My original reconstruction (Fig. 19) had one small problem: the animal was upside-down. This became

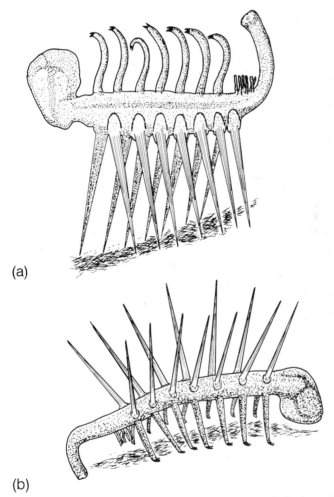

(a)

(b)

Fig. 19. (a) Original reconstruction of *Hallucigenia sparsa*, with the animal depicted as walking on its seven pairs of stilt-like spines. (b) Revised reconstruction, with the spines now forming a defensive array above the animal.

apparent only when somewhat similar animals were discovered in a new fossil site in China (in Yunnan Province), known as Chengjiang. As we shall see, preservation similar to the Burgess Shale is now known to occur quite widely, although Chengjiang and Sirius Passet (in North Greenland) are the most important of the new discoveries (Fig. 13). The related animals from Chengjiang, studied by a Chinese and a Swedish palaeontologist (Hou Xianguang and Lars Ramsköld[27] respectively) showed that when alive *Hallucigenia* walked on a set of tentacle-like legs and protected itself with a ferocious palisade of spines (Fig. 19b). While I had thought that there was only a single set of tentacles, these workers were able to demonstrate a second set. *Hallucigenia* still looks a very strange animal, but as we shall see it turns out in fact to be a sort of arthropod. So is *Hallucigenia* completely understood?

Not yet, because there is now disagreement about which is the head and which the tail.

The story of how sense was made of *Hallucigenia* is comparatively straight forward. The history of the interpretation of *Anomalocaris* is rather more complex. As explained above, this fossil was found first in the *Ogygopsis* Shale (Fig. 11). It does occur in the Burgess Shale, but is much more abundant in the former deposit. The palaeontologist Whiteaves[28] thought it must represent the abdomen of a primitive crustacean, equivalent to the 'tail' of a prawn which one pulls off to obtain the succulent white meat. When Derek Briggs started his reinvestigation of the Burgess Shale he quickly realized that the idea of this fossil representing an abdomen was extremely unlikely. In fact the only reasonable interpretation for these fossils was as some sort of limb, most probably being used for walking.[29] The fossils of *Anomalocaris* (Fig. 11) consisted of a series of units, articulating with each other and having the same basic design as the leg of an arthropod, such as occurs in flies and lobsters. The only snag with this idea was that the specimens of *Anomalocaris* were conspicuous and large, often reaching 10 centimetres or so in length. If the legs were that size, the implication was that the original animal must have been quite a lot bigger. Not unreasonably Derek Briggs proposed that the original animal, although still otherwise unknown, had a row of such legs arising along either side. Thus it would have looked somewhat like a centipede, but would have been much larger. An approximate calculation suggested a possible length of a metre. The difficulty in testing this idea was that no complete fossils had been found. The limbs were always found isolated, and not obviously attached to any other fossil. Despite these problems everybody was happy that *Anomalocaris* had to be some sort of arthropod.

The thread of the story now takes an unexpected turn. One of the curiosities of the Burgess Shale is a general absence of jellyfish;[30] curious because jellyfish are very abundant today yet are thought to belong to a primitive group of animals, the cnidarians, which should have evolved before the Cambrian. Nevertheless, Charles Walcott thought he had discovered a jellyfish. He named it *Peytoia*, but it was certainly rather peculiar (Fig. 20). Overall it looked rather like a pineapple ring, with the centre cut out. The body of this fossil was composed of a series of plate-like structures that surrounded the central hollow. In total there were 32 of these plates. Four large ones occupied each of the corners of *Peytoia*, and they were each separated by seven narrower plates. Another peculiarity was that the plates had prong-like structures that projected into the central opening. *Peytoia* certainly did not look like any other jellyfish, but who was to say what strange variants might have emerged during the Cambrian explosion?[31]

While I was looking through the collections in the Smithsonian Institution I noticed a very odd specimen. Walcott had already described it, believing it to be some sort of sea cucumber (zoologically they are known as holothurians,

Fig. 20. The mouth-parts of *Anomalocaris*, originally interpreted as the jellyfish *Peytoia*.

and are relatives of the starfish and sea urchins). At one end of this fossil was a circular structure, and quite plainly this was identical to *Peytoia*. The remainder of the specimen was rather poorly preserved, but Walcott's idea that it was a sea cucumber seemed rather far-fetched. The conclusion I came to was perfectly straightforward, but wrong. I suggested that the specimen represented a chance association: somehow a specimen of *Peytoia* had come to rest on top of another fossil which I thought was some sort of sponge.[32]

How and why I was mistaken became clear only when, back in Cambridge, Harry Whittington decided to have a closer look at a large fossil that the GSC had collected. This specimen had been puzzling him for some time. The specimen was neither particularly spectacular, nor was it complete, but it seemed to consist of an elongate body on either side of which arose a series of overlapping lobes. Harry Whittington started his excavation of the fossil. Before long, a new structure, previously hidden, appeared. Excavation continued and to Whittington's very considerable surprise, there lay a limb of *Anomalocaris*, but quite clearly it was attached to the front region of the animal (Fig. 21). But more was to follow. Further preparation of other specimens revealed a second sensation. Careful chipping of the overlying layers revealed that close to the limb there was a circular structure. Now there was

Fig. 21. The specimen of *Anomalocaris* originally prepared by Harry Whittington to reveal the giant anterior limb. [Photography courtesy of H.B. Whittington (University of Cambridge).]

no doubt. Here, in the same animal, was what everybody had previously thought to be the free-swimming jellyfish *Peytoia*.[33]

The route to understanding the *Anomalocaris* animal, therefore, was a catalogue of errors. What Derek Briggs had thought were the limbs of a giant centipede-like animal transpired to be a prominent pair of limbs at the front of the animal. What Charles Walcott had believed to be a jellyfish was now seen to be an integral part of the animal, forming an extraordinary mouth and jaw, which was apparently capable of holding and probably puncturing struggling prey. What I had interpreted as a composite fossil, formed by the chance association of a *Peytoia* and a sponge, was actually a poorly preserved specimen of the original animal. So if bits of *Anomalocaris* had been thought to be a jellyfish or a giant leg, what was the new verdict? Harry Whittington's new study, which he conducted jointly with Derek Briggs, showed that this was one of the most remarkable animals in the Cambrian seas. It probably grew up to a metre long. In addition to the anterior pair of limbs, which presumably were used to hold and manipulate prey, the body carried a series of flexible lobes. It was argued that their undulatory motion propelled the animal through the water, in search of food.

But is *Anomalocaris* completely understood? Almost certainly not. Quite recently a number of new specimens have been discovered. They reveal that at the posterior end there was spectacular tail fan, perhaps used to balance the animal while swimming.[34] And further surprises have emerged. Specimens from Chengjiang show that at least some species of animals closely related to *Anomalocaris* were equipped with a series of legs, situated beneath the lobes.[35] These legs would have enabled the animal to stroll across the sea floor. Even with these modifications, *Anomalocaris* still looks very extraordinary to our eyes. But fundamentally it is not that strange. As our knowledge of this animal continues to grow, so it is becoming clear that it is some sort of arthropod.

Work on the Burgess Shale still continues in Cambridge, but the main period of activity is now past. Lives and careers have moved on, and there are many new problems to tackle. As we shall see below, some of this new research is a direct extension of the Burgess Shale programme. This is especially true of our work on the Sirius Passet fauna from North Greenland. But elsewhere the Burgess Shale itself is still a focus of research interest. For a number of years Des Collins of the Royal Ontario Museum in Toronto has been pursuing vigorously a series of excavations in the vicinity of Field (Fig. 13). There have been extensive searches for new localities, and these have led to impressive discoveries on Mount Stephen and Mount Field. In addition, quarrying at several horizons close to the Walcott Quarry has yielded some marvellous new collections. Most of them have received only preliminary study, but some of the new information on *Anomalocaris* comes from newly discovered specimens. There is no doubt that more significant discoveries will be reported before too long.

Meanwhile, back in the MCZ in Harvard a young postgraduate, Nick Butterfield, was working on Proterozoic fossils. As part of his research he had to master very delicate methods of fossil preparation, including the digestion of rock by hydrofluoric acid. This chemical is an extremely corrosive reagent, but some organically preserved fossils are resistant to its attack and so can be isolated from the surrounding matrix and studied under the microscope. Nick Butterfield thought it would be interesting to see if he could extract fossils from the Burgess Shale using this technique. He succeeded, and obtained some very interesting results. In particular, he discovered that the microstructure of some fossils was preserved, a rather unexpected conclusion because it had been generally assumed that the fossils were squashed flat and replaced by films composed of silicate minerals.[36]

When the Cambridge team began their investigations 25 years ago nobody had an inkling of just how remarkable the new discoveries would turn out to be. The discovery of similar faunas elsewhere in the world is showing that the Burgess Shale is far from being unique, although the sheer quality of the fossil preservation and the extent of the collections still leave it unrivalled. The investigations into the newly discovered faunas are only just beginning, but I believe that in even less time than was taken by the Cambridge Campaign the future discoveries and interpretations will also lead to some profound reassessments. But for the moment at least the Burgess Shale remains the exemplar, the yardstick of Cambrian life. Our knowledge of this vanished world is effectively a mental construct; it is an exercise in scientific imagination. If this remarkable fauna is to be made accessible we must move on, but back in time. Now is the opportunity to transmogrify all those scientific monographs and technical papers, so that from the arcane pages of stolid scientific prose we may embark on a journey to the Cambrian period, to visit the Burgess Shale itself, some 530 million years ago. I want you to imagine that the art of time travel has been perfected and we are to accompany two zoologists back to the Cambrian in a specially designed vehicle, fully equipped to study the wealth of animals that await us.

Notes on Chapter 3

1. A gripping account of the building of this railroad is to be found in *The last spike: the great railway, 1881–1885* by P. Berton (McClelland and Stewart, Toronto, 1971).
2. To date there is no authoritative and detailed account of the general palaeontological work in the vicinity of Field, including the discoveries that led up to the uncovering of the Burgess Shale itself. Some of the main details are, however, given in a short article in *Rotunda (The Magazine of the Royal Ontario Museum,* Vol. 19, pp. 51–7 [1986]) by D. Collins.
3. This paper is to be found in *Canadian Record of Science* (Vol. 5, pp. 205–8 [1892]).
4. This, the first of C.D. Walcott's contributions to the palaeontology of this area, was published in the *American Journal of Science* (Vol. 36 (series 3), pp. 161–6 [1888]).
5. An account of this field season and the fossils collected is given by C.D. Walcott in the *Canadian Alpine Journal* (Vol. 1, pp. 232–48 [1908]).
6. This paper by H. Woodward, published in *Geological Magazine* (Vol. 9 (new series, decade 4), pp. 502–5 and 529–44), described some fossils collected on Mount Stephen by the mountaineer Edward Whymper, most famous for his earlier ascent of the Matterhorn, an expedition marred by tragedy.
7. The likely circumstances of the discovery of the Burgess Shale were discussed by S. J. Gould in his book *Wonderful life* (Norton, New York, 1989), and are also carefully considered by E.L. Yochelson in the *Proceedings of the American Philosophical Society* (Vol. 140, pp. 469–545 [1996]). Yochelson's thoughtful paper is well worth reading, not least because of his sympathetic and balanced portrayal of Charles Walcott.
8. These papers were incorporated into a series that Walcott entitled 'Cambrian Geology and Paleontology', and were published in *Smithsonian Miscellaneous Collections.*
9. This last paper by C.D. Walcott on the Burgess Shale appeared in *Smithsonian Miscellaneous Collections* (Vol. 85, pp. 1–46 [1931]).
10. These words appear in an annual report by C.D. Walcott, published in *Smithsonian Miscellaneous Collections* (Vol. 68, pp. 4–20 [1918]).
11. These included a strange organism *Petaloptyon danei*, described in the *Bulletin of the Museum of Comparative Zoology at Harvard College* (Vol. 55, pp. 165–213 [1931]), which appears not to have received further examination, at least in any detail.
12. P.E. Raymond's brief report of this excavation can be found in *Annual Report of the Museum of Comparative Zoology* (for 1929–1930, pp. 31–3 [1930]).

13. This hypothesis by N.J. Butterfield is to be found in *Lethaia* (Vol. 28, pp. 1–13 [1995]).
14. The most comprehensive treatment of the pyritized fossils of the Hunsrück Shale, which are amenable to radiography, is *Fossilien in Hunsrückschiefer: Dokumente des Meereslebens in Devon*, by C. Bartels and G. Brassel, published by Museum Idar-Oberstein (Vol. 7, pp. 1–231 [1990]).
15. The best available synthesis on the Mazon Creek fauna presently available is *Richardson's guide to the fossil fauna of Mazon Creek* edited by C.W. Shabica and A.A. Hay (Northeastern Illinois University Press, Chicago, 1997). See also the book edited by M.H. Nitecki entitled *Mazon Creek Fossils* (Academic Press, New York, 1979).
16. The book *Solnhofen: a study in Mesozoic palaeontology* (Cambridge University Press, 1990) by K.W. Barthel and others gives a useful overview.
17. The wonders of the Eocene oil-shales are graphically described in the book *Messel: an insight into the history of life and of the Earth* (Clarendon Press, 1992) edited by S. Schaal and W. Ziegler.
18. The details of H.B. Whittington's paper on *Marrella* may be found in *Bulletin of the Geological Survey of Canada* (Vol. 209, pp. 1–24 [1971]).
19. The study of *Sidneyia* by D.L. Bruton appeared in *Philosophical Transactions of the Royal Society of London* B (Vol. 295, pp. 619–56 [1981]).
20. C.P. Hughes' study of *Burgessia* was published in *Fossils and Strata* (Vol. 4, pp. 415–35 [1975]).
21. This key paper by H.B. Whittington was originally a contribution to the Symposium on Extraordinary Fossils held in Chicago, and published as Part I of the *Proceedings of the North American Paleontological Convention* (pp. 1170–201 [1971]).
22. But see note 13 of this Chapter, and also note 2 of Chapter 1.
23. The following section is also rehearsed in the chapter I contributed to *Cambridge minds* (Cambridge University Press, 1994).
24. The census of the Burgess Shale fossils, together with inferences on the community structure, was published by me in *Palaeontology* (Vol. 29, pp. 423–64 [1986]).
25. My paper was published in *Palaeontology* (Vol. 20, pp. 623–40 [1977]).
26. Richard Cowen's prescient view is discussed (pp. 83–4) in the second edition of his book *History of Life* (Blackwell Scientific Publications, Oxford, 1995).
27. The key paper is by L. Ramsköld and X. Hou in *Nature* (Vol. 351, pp. 225–8 [1991]). Ramsköld subsequently reiterated the point about *Hallucigenia* in *Lethaia* (Vol. 25, pp. 221–24 [1992]). Since then the description and interpretation of fossils related to *Hallucigenia*, which are notably abundant in Chengjiang, has flourished and a series of papers has appeared. Key items are the publications by L. Ramsköld (*Lethaia*, Vol. 25, pp. 443–60 [1992]), X. Hou and J. Bergström (*Zoological Journal of the Linnean Society*, Vol. 114, pp. 3–19 [1995]) and J.-Y. Chen and others (*Bulletin of the National Museum of Natural History, Taiwan*, Vol. 5, pp. 1–93 [1995]).
28. See note 3 of this chapter for details.
29. The redescription of the limbs of *Anomalocaris* by D.E.G. Briggs can be found in *Palaeontology* (Vol. 22, pp. 631–64 [1979]).
30. Des Collins, of the Royal Ontario Museum in Toronto, has told me that among his newly acquired collections are some possible examples of jellyfish.
31. The true nature of *Peytoia* emerged before there was any attempt to redescribe the Burgess Shale fossils as jellyfish. I had, however, described a specimen from the Middle Cambrian with R.A. Robison (*Journal of Paleontology*, Vol. 56, pp. 116–22 [1982]), as a jellyfish, albeit with the proviso that it looked very odd. This paper was significant in another respect inasmuch as it was one of the first to indicate that faunas of the Burgess Shale type were considerably more widespread than was thought.

32. The interested reader can find this paper in *Journal of Paleontology* (Vol. 52, pp. 126–31 [1978]).

33. This major reassessment of *Anomalocaris* was published by H.B. Whittington and D.E.G. Briggs in *Philosophical Transactions of the Royal Society of London B* (Vol. 309, pp. 569–609 [1985]).

34. This discovery is one item to emerge from a spectacular series of new discoveries by the Royal Ontario Museum's excavations at the Burgess Shale and nearby localities. The report on the new anomalocarids is by D. Collins in *Journal of Paleontology* (Vol. 70, pp. 280–93 [1996]).

35. This paper, by X-G. Hou and others, is published in *Geologiska Föreningens i Stockholm Förhandlingar (GFF)* (Vol. 117, pp. 163–83 [1995]).

36. The description of this technique and some preliminary conclusions reached by N.J. Butterfield can be found in *Paleobiology* (Vol. 16, pp. 272–86 [1990]). Butterfield subsequently extended this work into deposits comparable to the Burgess Shale in North-West Canada, with most impressive results. The two key papers showing further examples of the extraordinary quality of fossil preservation, including microstructural detail, are to be found in *Nature* (Vol. 369, pp. 477–9 [1994]) and (with C.J. Nicholas) in *Journal of Paleontology* (Vol. 70, pp. 893–9 [1996]).

CHAPTER 4

Journey to the Burgess Shale

Departure

Imagine that time travel was really possible. If a machine capable of travelling into the past was ever invented, then human curiosity would know no bounds. Instead of working with bits of bones and pieces of broken shell, palaeontologists would be able to visit ancient seas and continents. There they would see the life of the past in all its strangeness, complexity, and richness. Let us suspend disbelief, ignore the laws of physics, and pretend that time travel is a reality. Picture in your imagination the scene. In the grounds of a special laboratory the time machine is parked. A ramp leads up to a door, which is open. Two scientists have just walked up the ramp and through the door. They are now completing the final check before departure. The time dial has been set for 520 million years in the past;[1] the destination is central Canada. Let us imagine that we have joined them. The door slides shut. At first the machine remains stationary, and then to the watching ground crew it begins to flicker and suddenly disappears.

This will be a journey in imagination, but one based as firmly as possible on the interpretation of the scientific evidence. Let us suppose that this journey back to the Cambrian will take an hour. At such a rate of travel, time has been telescoped by about a thousand million times. One minute represents almost nine million years, every second is equivalent to almost 150,000 years. Human history as recorded by the development of agriculture, the invention of writing, and the building of the first cities takes a mere 50 milliseconds. On this scale our species, *Homo sapiens*, has existed for less than four seconds. The earliest hominids, represented by the australopithecines (Fig. 1), make their debut in Africa after 40 seconds of time travel. In the machine time continues to race backwards. Seven minutes have gone by. If we stopped now we would be in time to see the remains of a giant comet bombard the Earth. The planet is shrouded in thick dust clouds, formed by pulverized rock powder thrown up from the impact craters and soot particles from the burning forests. Huge tsunamis wash over the coastal regions. With sunlight blocked by the dust clouds the temperature has plummeted. On the continents the dinosaurs are dying in their millions; in the sea the food chains are collapsing. We would be witnessing the end-Cretaceous extinctions, which took place some 65 million years ago. But on this journey there is no time to stop. After 25 minutes of travel the first mammals appear, a little before the earliest dinosaurs. Passing the mass extinction at the end of the Permian period, we now hurtle back through the Palaeozoic. If the crew were to slow the machine down after 35 minutes we could make a choice. Hovering over what is now England, we would see a landscape covered by dense forests, luxurious plant growth flourishing in a hot and humid atmosphere. Debris from these plants accumulating in swamps as thick layers of peat will ultimately be transformed into coal. But if the machine were moved to Australia a very different scene would greet us and we would see glaciers snaking down to a sea dotted with large icebergs. Although England at this time is close to the equator, Australia is near to the South Pole and in the grip of an ice age. Further back in time the vegetation of the continents begins to change. Trees disappear, and after 50 minutes have elapsed the only plants visible are those that form a thin cover, restricted to wet bogs and the edges of streams. The end of the journey is in sight: in an hour we have travelled back to the Middle Cambrian.

The machine settles to the ground. We are about 300 km from the sea. The ramp unfolds, the door opens and we descend. What do we see? We have landed close to the equator and it is a warm morning, although the Sun is not quite as bright as today. Somehow our breath seems a bit short. A glance at the atmosphere monitor reveals why. Oxygen levels are somewhat lower than the present-day level of 21 per cent. The surrounding scenery is very bleak. The entire continent is a huge desert. A few kilometres away there are large sand dunes. The machine has come to rest on the edge of a gully covered with the anastomosing channels that fill with water only during occasional flash floods. Vegetation, however, is not entirely absent. Where moisture lingers small plants, similar to living mosses, survive. There are, however, no animals: no snakes, no birds, no insects.

The purpose of this trip is, of course, to visit the Burgess Shale. It will therefore be necessary to fly beyond the land, out towards the open ocean. Crossing the continent the land below remains desert, flat and monotonous. The sea appears on the horizon, and as the machine drifts over the shore we see a line of pounding surf stretching as far as the eye can see. Close to the coast the seabed is composed of sands and silts. These are derived, of course, from the erosion of the land and are constantly being stirred by tides and currents. Further offshore, however, the water becomes much clearer. One of us comments that the view is remarkably similar to what one sees from an aeroplane flying over the Caribbean, especially over the region known as the Bahama Banks. The blue water is quite shallow, and the sea floor can be seen to be composed of carbonate sediments arranged in shoals and banks. Ultimately this region will be transformed into huge thicknesses of limestone.[2]

The time machine now climbs higher into the sky. Behind it the land is only just visible in the distance. In fact, with North America straddling the equator in the Middle Cambrian, to the north of the continent there was no land all the way to the North Pole. Ahead of us, therefore, the sea stretches as far as the eye can see, but a few kilometres away there is a distinct line of surf. Beyond this line the colour of the water changes from a pale to a deep blue. We have reached one of the wonders of the Cambrian world, the Cathedral Escarpment (see Glossary). This is a huge submarine cliff running along the edge of the continental shelf. At the base of this cliff lie the sediment and fauna that will go to form the Burgess Shale.

In a few places along to the rim of the Escarpment there are small islands.[3] It is towards one of these that the time machine descends in a controlled glide. Once it has landed the final preparations are made to explore the Burgess Shale, using a small submersible vehicle. Apart from its visitors the island is devoid of life. A strong ocean swell is running in, but out to sea no dolphins or whales are visible. The sky is almost cloudless, but there are no seabirds.

The submersible descends

The submersible first has to travel just beyond the edge of the Escarpment; then, floating on the deeper water, it begins to dive. The seabed at the toe of the Escarpment lies at a depth of almost 100 m.[4] The submersible now sinks through the water, hugging the edge of the escarpment, which forms an almost vertical wall of limestone. At the base of this cliff, however, the sediments have a different composition. They consist of muds and silts, and it is on, in, and above this sea floor that the Burgess Shale animals live. At this depth the water is gloomy, for little of the tropical sunlight reaches this far down. The seabed, however, is teeming with animals. It is time to look at them in some detail.

How do animals live in the marine environment? Many inhabit the sediment, perhaps burrowing through it in search of food. These represent the infauna, what here we call the mud-dwellers. Others live on the seabed itself: this is the epifauna. Some of these animals are attached or rooted to the sea floor and so are capable of only very limited movement. They form the so-called sessile epifauna, termed here the mud-stickers. In contrast, the vagrant epifauna can walk or crawl across the seabed. They will be referred to as the strollers. By no means all animals, however, are confined to the seabed. Some are capable of swimming only short distances, but others live more or less permanently in the water, clear of the seabed. Active swimmers are known as the nekton. Finally, organisms that float or drift form the plankton. Unless equipped with structures such as gas-filled floats, most of the plankton are tiny because even quite small objects sink rapidly in water.

The divisions between these various categories are not clear-cut. Consider jellyfish. They swim by contractile pulses of the bell-shaped body, and in quiet water can propel themselves quite rapidly as swimmers. But more often they are at the mercy of tides and currents, as witnessed by the thousands that may be thrown up on a beach after a storm. Other animals may belong to the category of stroller, but in times of danger bury themselves in the sediment and so become temporary mud-dwellers until they dig themselves free. Ecologists now realize that the basic ecological structure of marine life has not changed radically since the Cambrian. Then as now it is possible to recognize without difficulty such categories as the mud-dwellers, strollers, or swimmers. When the actual faunas are compared, however, the differences are for the most part much more profound, although even half a billion years of evolution have failed to erase all differences. For example, in the Cambrian the dominant group of mud-dwellers were a group of worms known as the priapulids. Today they flourish only in a few selected areas, and for the most part their place has been taken by groups such as the polychaete annelids, and to a somewhat lesser extent some snails and clams, as well as heart urchins and other sea urchins. Similarly the strollers of the Cambrian included numerous arthropods, but for the most part they belong to groups very different from those found today, most of which are crustaceans. Not only that, but joining the strollers today we see a considerable variety of snails and echinoderms, especially the starfish, brittle-stars, and sea urchins. Perhaps the most profound differences between the seas of the Cambrian and today would be seen among the swimmers. Today the nekton is dominated by creatures such as the fish, marine mammals (whales, dolphins), and the squid. In the Cambrian none of these animals had evolved, and among the most active swimmers the arthropods were probably conspicuous. Even so, in this category similarities certainly exist. Today the ctenophores are a very important group, and much the same seems to have applied to the Cambrian faunas.

Let us return to the submersible. Imagine that it is equipped with a small laboratory, capable of capturing and studying the Burgess Shale animals. In reality, of course, all our information comes from the crushed and flattened fossils, and in some instances a species may be known only from a single specimen. What follows, therefore, is very much an exercise in scientific imagination. Not all of what I describe can be correct. It is important, however, that we try to bring ancient communities such as the Burgess Shale back to life. In this way it is much easier to enter ancient worlds, that at first sight appear to be very alien and remote from the one we inhabit.

Mud-dwellers

Picture, therefore, the submersible drifting slowly across the sea floor. Behind it the cliffs of the Cathedral Escarpment rear upwards towards the warm sunlit waters of the surface. The seabed, where the Burgess Shale animals live, slopes away from the Escarpment, but the surface is rather irregular with a hummocky surface and there are occasional scoop-like depressions. The first thing that the scientists in the submersible will investigate is what animals are living in the sediment itself. This infaunal community is dominated by a group of worms known as the priapulids. The group has persisted until today, but nevertheless living examples of priapulids (Fig. 22) are rather obscure.[5]

Fig. 22. An example of a living priapulid worm. [Reproduced from Fig. 1 of V. Storch, R.P. Higgins, V.V. Malakhov, and A.V. Adrianov. Microscopic anatomy and ultrastructure of the introvert of *Priapulus caudatus* and *P. tuberculatospinosus* (Priapulida). *Journal of Morphology*, Vol. 220, pp. 281–93. Copyright © (1994, John Wiley). Reprinted by permission of Wiley–Liss, Inc., a subsidiary of John Wiley & Sons, Inc.]

Certainly there are plenty of well-qualified zoologists who know almost nothing about them. Suppose that the submersible had a suction device that could be inserted into the sediment in search of these mud-dwellers. The first Burgess Shale worm to be recovered is the priapulid *Ottoia* (Fig. 23; also Plate 1).[6] It is

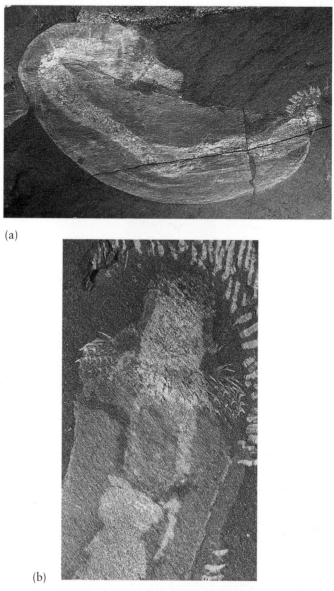

(a)

(b)

Fig. 23. (a) The Burgess Shale priapulid *Ottoia prolifica*. The anterior proboscis is bent around at the left-hand end. Note the prominent intestine. (b) Detail of anterior; note the hooks and at the anterior tip the teeth. The white marks around the specimen were produced by the dental drill used to remove overlying sediment.

an extremely active burrower; if released it will vanish almost immediately into the soft sediment. The fleshy-white body of *Ottoia* usually is about eight centimetres in length, although the largest specimens exceed 15 centimetres. The body is divided into two main regions. The anterior is composed of a structure known as the proboscis. This has the curious property of being able to be retracted into the front end of the body or alternatively being extended forwards. A simple analogy is to think of it as resembling the finger of a rubber glove. By pushing on the tip it can be folded inwards. The simplest way to evert it is to blow into the end of the glove and watch the finger pop back into its usual position.

The proboscis of a priapulid is notable for its hooks and spines. In *Ottoia* there are about 25 rows of hooks, pointing backwards. This region of the proboscis is somewhat bulbous, but it then becomes narrower. At the end of the proboscis there is the mouth. A glance at the mouth reveals a formidable battery of sharp teeth. Not only do they surround the mouth, ready to grasp prey, but the first part of the digestive tube is also lined with a dense coating of sharp teeth. These teeth project inwards and downwards. If the prey tries to struggle free when it is swallowed, it will impale itself against these teeth. It is a formidable and dangerous apparatus.

The rest of the body of *Ottoia* is composed of an elongate trunk. Its surface is divided into a series of annulations, so that the trunk appears to be segmented. At the posterior end of the trunk there is a prominent row of hooks. Each hook is stoutly built and curves forward. The specimens of *Ottoia* are semi-translucent and some of the internal organs are vaguely visible through the body wall.

The scientists select one of the specimens of *Ottoia*. First it has to be anaesthetized, to relax the muscles before being killed. Now the worm can be dissected to reveal its internal anatomy. The body wall is cut open with a scalpel. The outer surface is a tough cuticle, beneath which lie the muscles of the body wall. They enclose a spacious cavity, which is filled with a pale-yellowish fluid. The most obvious feature within the body cavity is the elongate intestine. This runs from the mouth, back along the length of the animal to the anus at the posterior tip. With the specimen pinned open on the dissection tray the other obvious features inside the body cavity are a series of thick muscular strands. They run between the inner wall of the proboscis and a position further back on the inside of the body wall and act as powerful retractor muscles. As their name suggests, when these muscles contract they withdraw the proboscis into the anterior of the trunk. How then is the pro-boscis re-everted? The method employed is actually very simple. *Ottoia* is basically a large sac, filled with body fluid. This fluid is effectively incom-pressible. Think of a syringe full of water, but with the nozzle blocked up. Now try and push the plunger. Normally the water will shoot out of the nozzle, but with no exit the incompressible water makes it impossible to squeeze the syringe plunger more than a few millimetres. Hence when the muscles of the body wall contract the pressure of the body fluid will rise. Something has to give way. With the retractor muscles relaxed, the proboscis promptly everts by the action of hydrostatic pressure.

Let us now see how *Ottoia* conducts its life. Part of its time is spent in a burrow, which is roughly U-shaped.[7] At other times the priapulid moves through the sediment, in search of new prey. Burrowing is very rapid and efficient. The worm uses a special sequence of actions that is known as the burrowing cycle. This process principally involves the proboscis. To start the cycle the animal everts the proboscis so that it pushes forward into the mud ahead. If the mud is fairly firm and resistant, however, there is a potential danger. As the worm pushes out its proboscis there will be an equal and opposite reaction so that the rest of the animal is simply pushed backwards. How does the animal prevent such a counter-productive move? The priapulid has to wedge itself against the sediment. This will minimize backward slippage as the proboscis probes forward. Wedging is achieved by a variety of mechanisms. The proboscis hooks, the posterior hooks, and the general curvature of the body all help to lock the worm into position. The extent to which the proboscis is able to evert is remarkable. When the proboscis is fully protruding, the entire anterior gut has also been turned inside out! Hence all the teeth, which normally line the first section of the gut behind the mouth, are now visible on the outside of the proboscis. Even more extraordinary, however, is that when completely everted the end of the proboscis swells into a large balloon-like structure. When inflated, this swollen end of the proboscis is able to grip the sediment. Muscles that act to shorten the length of the animal now contract. Anchored firmly at its anterior end, the worm is dragged forwards. As it does so the proboscis inverts, until the animal is ready to begin the next round of the burrowing cycle. As described here the whole process of moving through the sediment probably sounds very laborious. In life, however, *Ottoia* is a rapid and effective burrower, with the proboscis everting and retracting every few seconds.

The specimen that has been dissected contains only unidentifiable remains of the worm's last meal in its intestine. The formidable teeth suggest, however, that *Ottoia* was an efficient and dangerous predator. What did it hunt? On the surface of the sediment there are scattered examples of an animal known as a hyolith (Fig. 24). They have no common name. This is probably because they are an extinct group, with no close living relatives.[8] Hyoliths secrete an external shell of calcium carbonate, in the same manner as a snail or clam. The shell of a hyolith is composed of several distinct parts. The largest component forms a hollow cone, in which the animal lived. One side of this cone is flattened, and this rests on the sea floor. The open end of the cone can be closed off with a lid, also known as the operculum. In times of danger this lid slams shut to protect the animal living inside. How did the

Fig. 24. An example of a fossil hyolith. [Photography courtesy of the Smithsonian Institution (Dr D. Erwin).]

hyolith move across the sediment surface? On either side of the animal there projected a long curved strut. Each strut joined the animal where the lid and conical shell met. Muscles were attached to the end of the strut, and movements of the struts helped to lever the animal slowly forwards.

The hyoliths browse peacefully on the surface of the sediment, seeking out organic detritus. But now close to a group of these hyoliths the sediment has begun to move slightly. What happens next is very fast. Suddenly the snout of an *Ottoia* rears out of the sediment, a hyolith is seized and quickly swallowed (Plate 1). Three times the group is attacked before the *Ottoia* sinks back into the mud. Each time a hyolith is swallowed in the same fashion, the priapulid grabbing the front end first. Why does *Ottoia* choose to attack this way round? Consider the curved struts sticking out of the hyolith (Fig. 24). These struts point backwards, and if the *Ottoia* swallowed the hyolith from the other direction there would be a danger of the prey becoming jammed in the gut of the priapulid. *Ottoia* is a voracious predator. Sometimes it will emerge out of the sediment to seize shellfish known as brachiopods. But *Ottoia* also hunts for food as it burrows through the mud. It readily consumes soft-bodied prey, and will even attack weaker individuals of its own species. *Ottoia* the predator is also a cannibal.

The scientists now start to hunt for other types of priapulid. Although *Ottoia* is by far the most abundant, several other types can be collected. The basic similarity of the proboscis in all these worms supports their identification as priapulids. Overall, however, there is a very wide variation of form. Indeed, each species of Burgess Shale priapulid is as different as any other, and an examination of two more of the Burgess Shale priapulids will emphasize some of these differences.

First, there is *Louisella* (Fig. 25; Plate 1).[9] The scientists had considerable difficulty in capturing *Ottoia* because it burrows so rapidly. In contrast *Louisella* is much less active. The animal is dark red in colour and can grow to more than 30 centimetres in length. It lives in a long horizontal burrow. A specimen is pulled out of its burrow, and at once the same basic division of proboscis and trunk is recognizable. The teeth of *Louisella* are not identical to those of *Ottoia*, but they are prominent and well adapted to subdue struggling prey. In the area of the proboscis where *Ottoia* had prominent hooks, *Louisella* carries soft lobate structures (or papillae). When the scientists yank the worm out of its burrow, they notice that the walls of its home are shored up with thick layers of mucus. Evidently the papillae of the proboscis are responsible for secreting the copious quantities of mucus necessary to line the burrow walls.

The trunk is very elongate, and has two remarkable features. Running along part of its length are two feathery rows, composed of tightly packed tentacles. These are obviously respiratory structures and act as gills to supply the worm with oxygen. Why does it need such elaborate structures when its relative *Ottoia* has no such equivalent? The reason appears to be that *Louisella* spends most of its life in its narrow burrow, and unless fresh sea water is brought in the oxygen will soon run out. The animal irrigates its burrow by a regular series of undulations that travel along the trunk. As these waves pass along they drive water over the surface of the worm, bringing in oxygenated sea water to the gills and sweeping away the stale water, together with the animal's waste products. One of the scientists now picks up

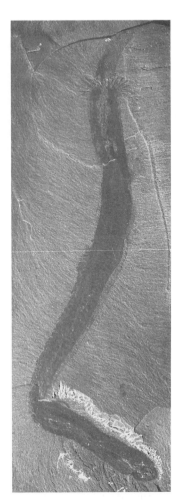

Fig. 25. The Burgess Shale priapulid *Louisella pedunculata*.

a captured specimen of *Louisella*. When handled it feels surprisingly rough. This is because on the trunk there are myriads of tiny stud-like spines, arranged in regular rows.[10] We are all puzzled about the possible function of these minute spines. After some discussion the scientists decide that they probably combine two functions. First, the spines provide some protection; and second they allow the animal to obtain a firm grip against its burrow wall.

In the initial survey of the seabed various skeletal remains scattered here and there were noticed. These include pieces of arthropod and mollusc shells. More puzzling are tubes, with a wall composed of a rather leathery organic material. The original occupant of these tubes, however, is soon discovered. It is another priapulid worm, known as *Selkirkia* (Fig. 26; Plate 1).[11] Most of

Fig. 26. The Burgess Shale priapulid *Selkirkia columbia*, showing the spiny proboscis projecting from the tube.

the animal is housed in the tube, which is open at either end. From the anterior end there protrudes a formidable spiny proboscis. It is this structure that betrays the priapulid affinities of *Selkirkia*. As the animal continues to grow so the tube must be extended. This is achieved by a special collar of secretory cells, located just behind the proboscis. These cells add tiny annular increments of new organic material to the front end of the tube. *Selkirkia* is also a burrower, but it lives head-down in the sediment. Only the posteriormost tip of the tube projects into the overlying water. Through this opening occurs the disposal of faeces and other waste products.

The Burgess Shale sediment teems with priapulid worms. This is in marked contrast to the seas of today, where in general priapulids are rare. Indeed, in present-day environments they often occur in fetid muds, stinking of hydrogen sulphide. This is an environment that is inimical to many other animals. Priapulids therefore seem to live in marginal environments. There is a strong suspicion that, although immensely successful in the Cambrian period, over geological time they were gradually pushed into the ecological sidelines by more efficient and successful competitors.

The scientists must now conclude their search for mud-dwellers. Their experience of present-day marine environments is that the most abundant type of worm belongs to the phylum known as the Annelida. To the non-specialist these are most familiar as the earthworms and the leeches. Not many people like to pick up these animals, especially because the leeches are famous as bloodsuckers. No story of an expedition to a tropical rainforest is complete without a graphic description of huge leeches looping through the undergrowth in search of an unexposed patch of explorer's skin. But if you can bear it, pick up an earthworm. Note first that the body is divided into a series of rings. These are the external indication of structures known as segments. Next gently run your finger along the surface of the earthworm. You should feel a slight roughness, but to see its cause you will need to look at the worm under a microscope. On each segment there are stout bristles. Each one is embedded in a pocket-like structure, and only the tips of the bristles project beyond the body. The bristles (or chaetae) are composed of chitin. This compound is also used to form the hard carapaces of animals such as beetles and centipedes.

Earthworms and their relatives mostly live in soils or fresh water. There are, however, marine relatives. These are very abundant and are known as polychaete annelids. Let us imagine that we are now looking at one of the Burgess Shale polychaetes, known as *Burgessochaeta* (Fig. 27; Plate 1).[12] It is quite small, seldom more than a few centimetres long. At the front end of the body there is a pair of prominent tentacles. The rest of the body is composed of about twenty segments. One segment is very much like another. What will be seen when a specimen of *Burgessochaeta* is placed under the microscope? Each segment carries four prominent bundles of chaetae, two on either side. Unlike the earthworms, where typically the chaetae are few in number and stout, in the polychaetes each bundle contains numerous, rather slender chaetae. The arrangement of two bundles of chaetae on each side of the polychaete body is also very characteristic. Those on the upper side are referred to as notochaetae, those towards the lower side form the neurochaetae. In many polychaetes the notochaetae differ in shape from the neurochaetae, but *Burgessochaeta* is unusual because the chaetae of each bundle are identical.

How did *Burgessochaeta* live and on what did it feed? Our scientists first noticed the worm by its anterior tentacles, which stretch from the opening of the burrow, feeling across the sea floor for particles of food. The burrow itself is vertical, rather like a chimney. *Burgessochaeta* climbs up and down

Fig. 27. The Burgess Shale polychaete *Burgessochaeta setigera*, anterior half only.

this burrow using its chaetae. They help to push the animal along, or if necessary wedge it against the burrow walls.

Mud-stickers

The next stage of the investigation is ready to begin. The submersible drifts slowly above the seabed in order to study the immobile epifauna, that is the mud-stickers. Particularly abundant are sponges. These are generally agreed to be the most primitive of animals. All are rooted to the sea floor and project upwards into the water. How do sponges feed? In the 1990s a living sponge was found living in the depths of the Atlantic that can entangle animals and then slowly digest them.[13] But this was an extraordinary and completely unexpected discovery, because all other sponges obtain their food by a method that is known as suspension feeding. This entails straining the sea water to capture minute food particles, including bacteria, by using a special scheme of filtration. In sponges the sides of the animal are perforated by tiny holes. It is through these pores that the sea water is sucked into the animal. Within the sponge the food is trapped using an ingenious mechanism. It is the beating of huge numbers of the hair-like cilia within the internal cavities of the sponge that set up the current necessary to drag in the sea water. Each cilium arises from a cell. Around the base of each cilium there is a collar-like structure. As the water passes through the collar, food particles are trapped and transported into the cell for digestion. Once the water has been filtered, it is then expelled from the sponge, jetting out from a large central opening.

As sponges are sessile they should in principle be very vulnerable to attack. Some Burgess Shale animals, notably the arthropod *Aysheaia*, may have lived on sponges. But in general sponges have devised a very effective means of protection. They secrete vast numbers of needle-like structures, made of either silica or calcium carbonate, that are known as spicules. These make a mouthful of sponges a most unattractive prospect. The spicules also have other functions, notably to help in supporting the body.

The submersible continues its slow cruise. The variety of sponges to be seen is remarkable.[14] Some are very small and stick up into the sea water by only about a centimetre. Others, however, reach quite substantial sizes, with a height of more than 20 centimetres. The most common of the sponges is called *Vauxia gracilenta* (Fig. 28; Plates 1 and 2). The first name (the genus) is in honour of Walcott's second wife Mary Vaux. The second name refers to its graceful form, a bush-like mass of slender, tubular branches. Unlike many of the other Burgess Shale sponges *Vauxia* does not have prominent mineralized spicules. This sponge supports and reinforces itself using a tough network of organic fibres.

Other sponges that are present are, however, very striking because of the prominence of their spicules. *Choia* is like a pincushion with elongate spicules radiating out in all directions (Fig. 29; Plate 2). This sponge is gregarious and several dozen individuals may occupy a small patch of sea floor. The sponge *Pirania* (Fig. 30; Plates 2 and 4) is more scattered in its distribution. It is more upright and is composed of several stout branches. Elongate spicules project outwards and upwards, to provide effective deterrence from attack. The scientists have scooped a specimen of *Pirania* off the sea floor. Now that it can be examined closely within the submersible they can see that the sponge has some guests. Attached to projecting spicules are a number of small brachiopods (Plate 4). They are a kind of shellfish with two valves enclosing the soft parts.[15] Like the sponges, brachiopods are also suspension-feeders. These animals filter the sea water using a complicated set of tentacles. (This feeding apparatus is known as the lophophore.) By settling on the sponge the brachiopods in effect are obtaining a free lunch. Not only do they benefit from the water currents set up by the much larger sponge, but perched above the sea floor the brachiopods are in no danger of sinking into the soft

Fig. 28. The Burgess Shale sponge *Vauxia gracilenta*. [Photography courtesy of the Smithsonian Institution (Dr D. Erwin).]

Fig. 29. The Burgess Shale sponge *Choia carteri*. [Photography courtesy of the Smithsonian Institution (Dr D. Erwin).]

muds of the seabed, and surrounded by the spicules of the sponge they are better protected against predators. Indeed, the soft nature of the sea floor can make its colonization by the mud-stickers quite difficult, especially if it has a soupy consistency. A hard object on the sea floor may thus provide a valuable point for colonization. One sponge is found attached to empty tubes of the priapulid *Selkirkia*. This is the globe-like *Eiffelia*, notable for its spicules forming star-like clusters (Fig. 31).

Although sponge spicules occur in a wide variety of shapes, they are all built in the same basic fashion. This is what makes the group known as the chancelloriids[16] so puzzling (Fig. 32; Plates 1 and 2). They are certainly very striking objects on the seabed, looking like large inverted cones and standing up to 50 centimetres in height. The outer skin is tough and wrinkled. Embedded in the skin are prominent spicules, consisting of several spinose

Fig. 30. The Burgess Shale sponge *Pirania muricata*. [Photography courtesy of the Smithsonian Institution (Dr D. Erwin).]

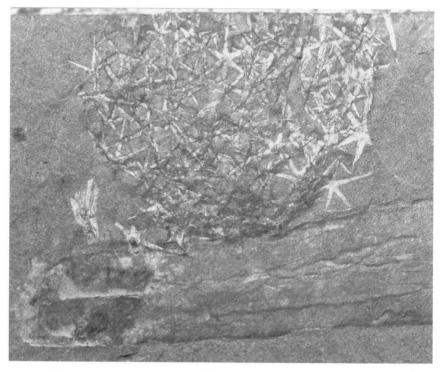

Fig. 31. The Burgess Shale sponge *Eiffelia globosa*. Note that it has attached itself to an empty tube of *Selkirkia* (see Fig. 26).

Fig. 32. The Burgess Shale sponge-like animal *Chancelloria*. [Photography courtesy of the Smithsonian Institution (Dr D. Erwin).]

structures radiating from a common base. In general, the chancelloriids are very sponge-like, but the spicules are constructed in a different way.[17] Unlike sponge spicules, which are usually solid or only have a very narrow internal cavity, in chancelloriids the interior of each spicule has a spacious cavity. This is filled with soft tissue that has a direct connection with the rest of the animal via a little pore at the base of the spicule. The scientists look carefully at the chancelloriids, but until the tissue can be returned to the laboratory for more detailed analysis they remain uncertain whether they are true sponges.

The submersible needs to cover a substantial area of sea floor, moving into slightly deeper water, before it encounters one of the rarest of the Burgess

Fig. 33. The Burgess Shale pennatulacean *Thaumaptilon walcotti*.

Shale animals. But the search is well worth while. It is one of the most important discoveries on the trip because it provides a direct link with the Ediacaran
assemblages (Chapter 2). The submersible has to approach very cautiously,
because at the slightest sign of danger the animal will shoot back into the sediment, retreating into its burrow. In the distance the animal looks like an
enormous frond or feather.[18] Indeed its name, *Thaumaptilon*, is derived from
the Greek words for 'wonderful feather'. The animal (Fig. 33; Plate 2) is
embedded in the sediment by a stout stalk. The rest of the body looks like a
sort of giant tongue. Overall it is about 20 centimetres long and pale red in
colour. One side of this tongue-like structure is more or less flat, but the
opposite side bears a series of branches that arise on either side of a central
axis. On each branch there are dozens of tiny star-like structures. In fact,
although *Thaumaptilon* looks like a single animal it is a colony. Each of the
'stars' is a minute animal, having the same basic structure as a sea anemone.

On close inspection the 'star' is seen to consist of a ring of tentacles that surround the mouth. As the water currents sweep over the frond the tentacles capture food. In each tentacle there are special cells (known as nematocysts or cnidae) that in contact with prey catapult a sting armed with poison. Exactly the same mechanism operates in the tentacles of jellyfish, which are relatives of this animal. In *Thaumaptilon*, however, the tentacles are too small to capture anything more than microscopic prey. Once the food has been stuffed into the mouth it is digested and a complex system of internal canals then distributes the nutrients to all parts of the animal.

Suppose that we had not stopped in the Cambrian, but had continued back for another 5½ minutes, equivalent to a time of about 580 million years ago. By doing so we would be able to study the Ediacaran assemblages.[19] We would then immediately be profoundly impressed by the differences between life in Ediacaran and Burgess Shale times, although the latter assemblage is far more diverse, much richer in form, and with a far more complex ecology. There would, however, be a striking point of contact via *Thaumaptilon*. This is because in the Ediacaran assemblages there are a wide variety of frond-like fossils. One of these, known as *Charniodiscus* (Fig. 8),[20] is astonishingly similar to *Thaumaptilon*. Quite clearly this Burgess Shale fossil is a relic from the past, an Ediacaran survivor.

The submersible now digs out a specimen of *Thaumaptilon*. The section stuck in the sea floor is slightly swollen. One of the scientists remarks that here is an obvious contrast with *Charniodiscus*, because in this Ediacaran animal the holdfast forms a prominent disc. They discuss why there is this difference. The answer seems to be that since Ediacaran times the world has been getting a more dangerous place. In the Ediacaran seas there were almost no predators. The disc of *Charniodiscus* was an ideal holdfast, well adapted to anchor the frond in sediments shifted by currents and tides (Fig. 8). By Burgess Shale times predators are much more numerous. It is now more important to be able to retreat rapidly, when necessary, into the safety of the seabed. Any sea pen with a broad, discoidal holdfast will find it almost impossible to retreat in this fashion, and the holdfast has to be modified in shape to allow easy withdrawal into its burrow.[21] Here predators are evidently driving the pace of evolution.

Are there any other stragglers from the Ediacaran world? The scientists in the submersible have discovered some curious animals. They are off-white in colour and look like elongate balloons, tethered to the sea floor. The name of this animal is *Mackenzia* (Fig. 34; Plates 1 and 2).[22] The body is tough and rubbery, but it can alter its shape from a squat, compact object to one that is much more elongate. The mouth is located at the top of the animal and it leads into a spacious digestive cavity. This cavity is lined with vertical partitions that principally serve to increase the surface area available for the absorption of nutrients. Only the lowest part of *Mackenzia* is embedded in

Fig. 34. The Burgess Shale animal *Mackenzia costalis*.

the sediment. To help secure itself the animal commonly gathers together skeletal fragments from the surrounding sediment, which are used to form a kind of hold-fast. In the Ediacaran faunas there are a number of bag-like organisms that are fairly similar to *Mackenzia*. Although this animal looks very different from *Thaumaptilon*, they are in fact, quite closely related. The sea pens, to which *Thaumaptilon* belongs, are placed in a major group of cnidarians known as the anthozoans (the word comes from the Greek and means literally 'flower-animal'). More familiar anthozoans, perhaps, are the sea anemones and corals, and it is to these animals that *Mackenzia* is fairly closely related.

Individuals of *Mackenzia* can grow to quite appreciable sizes, at least 15 centimetres high. The last member of the mud-stickers to be discovered on this expedition is however, much smaller. It projects only a centimetre or so above the sea floor, and is a beautiful animal, known as *Dinomischus*

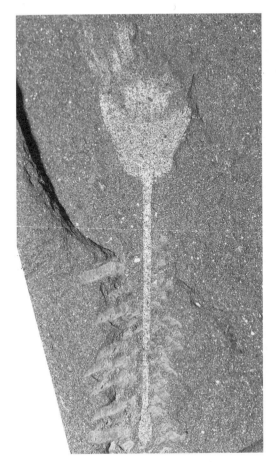

Fig. 35. The Burgess Shale animal *Dinomischus isolatus*.

(Fig. 35; Plates 1, 3 and 4).[23] From a distance it looks rather like a flower, perhaps a daisy on a long stalk. The body, perched above the sea floor, is goblet-shaped. On its upper rim it bears a palisade of plates, each with a pointed end. These plates are covered with cilia. The beating of these cilia strain the sea water for food, which is then transported towards the mouth, located within the circlet of plates. The stalk is elongate and very slender. By careful digging a specimen of *Dinomischus* is loosened from the mud. The end of the stalk is now visible and is revealed as a slightly swollen holdfast. Such an arrangement, with a cup-like body perched on an elongate stalk, has evolved in a wide variety of animals and is not of particular evolutionary importance. The more closely the scientists look at *Dinomischus*, the more puzzled they are about its relationships to other groups. Here is one of several enigmas in the Burgess Shale.

Strollers, walkers, and crawlers

Even though all of us in the submersible have been busy examining the mud-stickers, we have noticed that many other animals are scurrying around over the sea floor. By far the most abundant is the arthropod *Marrella* (Fig. 16; Plate 3).[24] Almost wherever we look several specimens are visible. It is a small animal, seldom more than a couple of centimetres long. *Marrella* looks strangely skeletal, but this is because the top part of the body consists a rather extraordinary set of spines. In detail, the anterior end is covered by a sort of box-like shield from which spring two pairs of spines, outer and inner, that sweep back over the body. The animals are moving rapidly over the sediment, but once one is captured it is possible to study it more carefully. Behind the head there extends a narrow segmented body. Each segment also carries a pair of appendages, which consist of two branches that arise from a common base (Fig. 36). The lower of the branches forms a walking leg. Although it is very small, only a few millimetres long, it is worth looking at closely. This is because the way in which the leg is built reveals one of the most characteristic features of an arthropod. Examined under the microscope the leg can be seen to be composed of a series of short cylindrical elements, which form an articulated series. The leg thus forms what is known as a jointed appendage. Indeed this basic arrangement extends to many other parts of a typical arthropod. The skeleton is external (an exoskeleton) and consists of rigid units which articulate with others via more flexible areas. It is a very ingenious design because the rigid skeleton confers protection for the internal soft parts, while the narrow articulating joints allow considerable

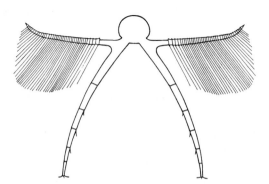

Fig. 36. Reconstruction of the biramous appendage of *Marrella splendens*. [Reproduced with permission of the Geological Survey of Canada, from text-fig. 8 (left) in H.B. Whittington (1971). Redescription of *Marrella splendens* (Trilobitoidea) from the Burgess Shale, Middle Cambrian, British Columbia. *Bulletin of the Geological Survey of Canada*, Vol. 209, pp. 1–24.]

freedom of action. In an attempt to find a moderately familiar comparison, zoologists usually compare the arrangement of the skeleton in arthropods to the armour of a medieval knight.

Let us return to the walking leg of *Marrella*. Although the skeleton is tough and rigid, the leg itself is remarkably flexible because it is composed of about six articulating units. Thus the leg can bend tightly or extend out straight. The skeleton of the leg, and indeed the rest of *Marrella*, is composed of the chemical known as chitin. A fly or a shrimp, both being arthropods, also have chitinous skeletons. The interior of each leg is hollow and largely filled with blood. Most of the blood in *Marrella* and other arthropods flows in large, open cavities rather than in narrow tubes such as our arteries. The space within the leg, however, also houses some muscles. They run between the joints, and by carefully controlled contraction can bend the leg into a variety of configurations. Of course, for walking the entire leg also needs to be moved. This is achieved by muscles that run between the near-end of the leg and the main part of the body.

The body of *Marrella* has about 20 segments, each with a pair of walking legs. Their movement is controlled from the central nervous system and is achieved by waves of activity running forward along the body. Each wave of movement affects about five pairs of legs. It is worth taking a close look at the precise mode of action of the legs. In succession, each pair is placed on the seabed and then swung back. The net result, of course, is that there is an equal and opposite reaction so that the animal moves forward. Once the leg has pushed back as far it can go, it is lifted up again and the process is repeated. At any one time about ten pairs of the walking legs are involved with the actual propulsion, while the remainder are being lifted clear of the seabed. It is also a very stable system, because at any one time only a few legs are clear of the ground and most are pressed against the ground. The sea floor is not perfectly flat, but highly irregular; so not only must the leg be swung into the correct position, but it must adjust itself to any topographic irregularities. As described, the operation of the mechanism sounds very laborious, and the power of each leg is relatively small. In the living *Marrella*, however, where 40 legs are acting together in rapid waves of movement, it is a very effective means of loco-motion. A little more consideration shows, however, that it is remarkably complex. Just imagine having to design a robotic system, not only to achieve the same pattern of action but in a structure only a few millimetres long!

What about the other branch of the appendage? This lies above the walking leg and is a feathery-looking structure (Fig. 36). In detail, it consists of an elongate shaft, projecting outwards and slightly backwards, that is made of a whole series of tiny cylinders. This shaft gives rise to a whole series of very fine, slender bristles. As the leg swings backwards and forwards, so do the filamentous branches. As they sweep through the water, these branches absorb oxygen and so act as gills.

Marrella shows many of the fundamental features of arthropods, including a segmented body with a series of jointed appendages. A peculiarity of *Marrella* is that along the length of the body the appendages are almost identical. This suggests that *Marrella* is rather primitive: in many arthropods, appendages in various regions of the body show different specializations and different functions. The simple arrangement, however, does not persist at the front end of *Marrella* (Fig. 16). Here, there are two specialized appendages, each consisting only of a single branch. The anterior pair are long and whip-like. These are the antennae. They are ceaselessly mobile, bending and straightening as they sense the environment into which the animal is moving. In *Marrella* these antennae are especially important because, unlike many other of the Burgess Shale arthropods, it is blind and has no eyes. Immediately behind the antennae is a pair of rather more robust appendages. These too are very active, but instead of whipping through the water they are regularly brushed over the seabed. On the edge of these appendages is a hairy fringe, ideal for trapping particles of food in the sediment. After several sweeps these feeding appendages are folded inwards, the hirsute margin is drawn across the mouth, and the food scraped off.

The Burgess Shale fauna is crowded with a remarkable array of different types of arthropods. The submersible gives us only limited time, and much that we would like to study must be left for a return trip. Before they departed for the Cambrian world the one type of arthropod which the scientists knew they were certain to see were the trilobites. They are dominant among Cambrian fossils, in part because they were extremely abundant. More significant, however, is that the arthropod skeleton is reinforced with calcium carbonate, specifically the mineral calcite. This means that the skeleton has a very good chance of becoming fossilized, because unlike the soft tissues the calcite is resistant to decay. Not all the exoskeleton, however, is thus reinforced. The underside of the trilobite, which includes the appendages, remains uncalcified. As a consequence it is very rare indeed to find these appendages fossilized.

The Burgess Shale houses a large number of trilobites. Most are rather small, a few centimetres in length, but a few are appreciably larger. It does not take the occupants of submersible long to discover a specimen of *Olenoides* (Fig. 37; Plate 2).[25] Viewed from above it is a typical Cambrian trilobite. The skeleton is divided into three main sections. At the front there is a head-shield (technically the cephalon); a somewhat similar plate covers the posterior end (the pygidium). In between there is a series of narrower segments, which in *Olenoides* total seven. Each of these segments is divided into three units, hence the name *tri*-lobite. There is an arched central axis, which is flanked by flatter extensions. At both ends of *Olenoides* there extend beyond the skeleton a pair of long slender appendages. Those at the front, as in *Marrella*, help to sense the environment in advance of the trilobite. Those at the back provide an early warning system against unexpected attack.

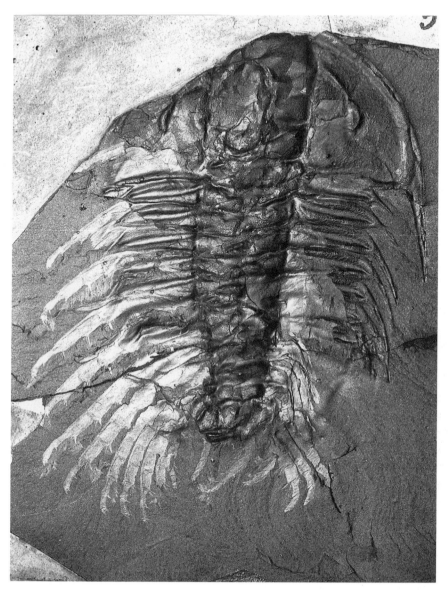

Fig. 37. The Burgess Shale trilobite *Olenoides serratus*. [Photograph courtesy of H.B. Whittington (University of Cambridge).]

The specimen of *Olenoides* is now seized by a special claw extending from the submersible. Once it is turned upside down and the long row of appendages is revealed. Although this trilobite is much larger, the basic construction of its appendages is quite similar to that of *Marrella*. Thus there are powerful walking legs, and above them the gills. Both the branches arise from a common basal unit (known as the coxa). Each coxa is attached to the body wall, and together they form a long series on either side of the midline, which defines a deep gutter. On closer inspection there appears an important difference from *Marrella*, because each coxa bears a set of ferocious spines. How they are used in feeding soon becomes clear. The trilobite is released and it quickly scuttles across the floor of the observation tank. One of the scientists then takes a small worm and places it in front of the trilobite. The antennae quickly sense it and the trilobite lunges forward. Some of the walking legs coil round the struggling worm and push it towards the central gutter, lined with the projecting spines. The prey is now helpless. Held and punctured by the spines it is passed forwards and shoved into the mouth, which faces backwards to receive food. The *Olenoides* is released and it scuttles off into the gloom, in search of more prey. Before it disappears it strides across a patch of firmer mud. As the legs move over the seabed, each leaves a small depression so that behind the trilobite there extends its walking trail.[26] Thus for the palaeontologist not only is there knowledge from the skeleton, but also from the traces it imprinted on ancient sea floors.

Over the millions of years trilobites have evolved a remarkable range of forms that reflect a multitude of ecological conditions. Some burrowed in soft sediment, others strolled across the seabed, and some swam near the surface of the oceans. There is, nevertheless, a basic similarity to all trilobites. The scientists therefore experience considerable surprise when they encounter the next arthropod, which is known as *Naraoia* (Fig. 38).[27] Once a specimen has been captured its appendages can be studied. They are remarkably similar to those of *Olenoides*, consisting of stout walking legs and prominent gills, both joined to the basal coxa, which is armed with spines. The scientists are surprised because the dorsal skeleton is not at all like a typical trilobite. In *Naraoia* it consists of two large shields, with a prominent zone of articulation running across the animal. Evidently what has happened is that in ancestors of *Naraoia* the skeleton formed a long series of segments, broadly similar to the arrangement in trilobites. These segments, however, have fused together to form the two shields. What has been sacrificed in terms of skeletal flexibility has led perhaps to improved protection and greater ease in pushing through the top layers of mud in search of prey.

There is another significant difference between *Naraoia* and trilobites. Although the skeleton of *Naraoia* is tough and resistant, unlike that of the trilobites it lacks impregnation by calcite. Without mineralization the carapace of *Naraoia* is translucent, and some of the internal organs are just discernible. A captured specimen is now killed in order to perform a dissection. The scalpel cuts into the anterior shield, and a section is then lifted away to reveal the internal organs. The anterior section of the gut is visible, and particularly notable is a bush-like mass of tubules that extend from the gut. These structures, known as the caeca, help to secrete digestive enzymes and also to absorb the food.

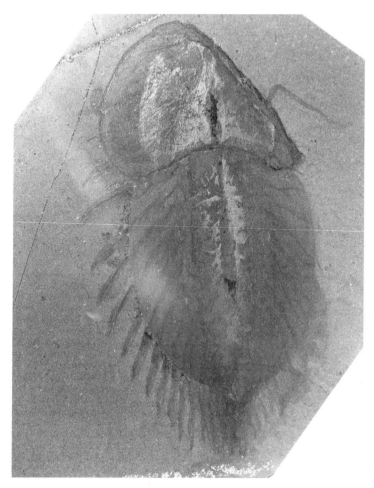

Fig. 38. The Burgess Shale trilobite *Naraoia compacta*. [Photograph courtesy of H.B. Whittington (University of Cambridge).]

So far all the arthropods examined by the time travellers have been active and mobile animals. Unless approached very carefully they will quickly scuttle off into the gloom. But other types of arthropod are somewhat more sluggish. Allowing the submersible to continue to drift across the seabed, the travellers encounter their first specimen of *Aysheaia* (Fig. 39; Plates 2 and 3),[28] walking slowly across the mud. At first sight, it looks more like a giant caterpillar. The tubercles, however, are arranged in regular rings around the trunk. The long worm-like trunk is supported above the seabed by a series of short, stubby legs, each of which ends in a couple of sharp claws. These legs are known as lobopods, because they are soft and lobe-like. When *Aysheaia* walks, waves of movement of the paired limbs pass forward along the body, each pair of lobopods pushing backwards and swinging forwards in the same

Fig. 39. The Burgess Shale lobopodian *Aysheaia pedunculata*. [Photograph courtesy of H.B. Whittington (University of Cambridge).]

manner as the walking legs of *Marrella*. There is, however, an obvious difference between these legs and the lobopods of *Aysheaia*. Although each lobopod is divided into a series of transverse rings, these are superficial and they do not represent the series of joints that make up the leg of an arthropod like *Marrella*. If a lobopod of *Aysheaia* were cut open it would be seen to have a rather thick, muscular wall that surrounds a cavity. In life this cavity is filled with blood. The action of the muscles and this fluid-filled cavity act together to provide a versatile hydraulic system that makes the lobopod an efficient and effective instrument for propulsion.

The time travellers watch the *Aysheaia* continue its march across the sea floor, as it makes its way towards a clump of large sponges. Now it begins to climb the sponge, easily maintaining its grip on the surface of the sponge by using the claws on the lobopods. The anterior end of the *Aysheaia* continues to move slowly from side to side, and then after a pause it bends down towards the surface of the sponge (Plate 2). The scientists have already noticed that the mouth of the animal is situated at the front end, and is surrounded by a series of forward projecting prongs. *Aysheaia* begins to feed on the sponge, lacerating its surface and sucking up small pieces of tissue. Having grazed on one region of the sponge, the animal slowly crawls forward and resumes feeding.[29]

While nestled among the branches of the sponge, *Aysheaia* has some degree of protection. But when it is walking across the sea floor how does it defend itself? It has no hard parts to provide a shield, and its soft, flexible body looks almost defenceless. Nevertheless, *Aysheaia* is well able to defend itself. The skin secretes a powerful and distasteful toxin that swiftly deters any would-be predator.

Shortly before embarking on the voyage of the time machine one of the scientists had been on an expedition to the tropical rainforest in Venezuela. Although this time traveller is now in a submersible beneath the waves of a Cambrian ocean, half a billion years in the past, the animal *Aysheaia* looks

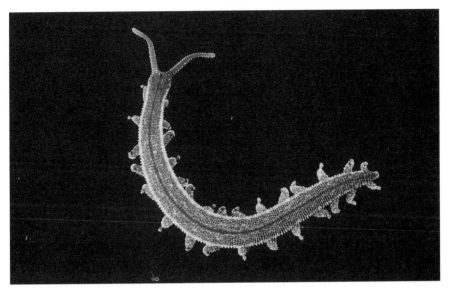

Fig. 40. The living velvet-worm, belonging to the group known as the onycho-phores. [Photograph courtesy of Amanda L. Reid (formerly ANU, Canberra). Photographed by Jenny Norman (Macquarie University, Sydney).]

strangely familiar. Then he remembers why this is so. During the expedition to Venezuela, one of the zoologists was collecting specimens of an animal known as the velvet-worm (Fig. 40), which includes the genus *Peripatus*. The Cambrian *Aysheaia* and living *Peripatus* look remarkably similar, most obviously in terms of the stubby lobopods arising from the caterpillar-like body. The animals, however, are not identical. *Aysheaia* is marine, whereas *Peripatus* and its relatives live only on land, typically in the leaf-litter and rotting logs of the forest floor. Another notable difference is that unlike *Aysheaia* the *Peripatus* has a powerful jaw, the teeth of which slice up prey. Actually this difference is not as profound as it first appears. These teeth derive from the claws of the first pair of lobopods, the size of which has become greatly reduced.

It has long been recognized that *Peripatus* (which belongs to a group known as the onychophores) is an arthropod. At first sight this may seem rather surprising because the onychophores do not have the characteristic jointed skeleton that we have already encountered in typical arthropods such as *Marrella, Olenoides*, and *Naraoia*. Nevertheless, *Peripatus* does possess a number of features that clearly demonstrate its place in the arthropods. For example, like other arthropods *Peripatus* (and *Aysheaia*) have a large body cavity that is filled with blood, which is why this cavity is technically known as the haemocoel. In fact, the lobopod-bearing arthropods such as *Aysheaia* and its living descendant *Peripatus* are exceptionally important in helping to understand the early evolution of arthropods.[30] This is because it is generally agreed that the jointed appendage evolved from some sort of lobopod limb. This transition was achieved by the hardening of the exterior (this process is known as sclerotization) so that ultimately the limb is encased in a skeleton that is divided into a series of units that articulate with each other by joints composed of flexible membranes. Why did this transition occur? The lobopod arrangement cannot be totally unsatisfactory; after all, it is still found in the living *Peripatus*. The principal reason, perhaps, is that the jointed appendage offers a far greater versatility of possible functions, with a whole range of specializations for locomotion and feeding. In addition, the way in which the jointed appendage is constructed provides very precise control and manipulation.

Does *Aysheaia* have any relatives in the Cambrian seas? The next animal to be encountered looks extremely strange. Indeed, when it is first spotted out of a window of the submersible we all burst out laughing. Here is *Hallucigenia* (Fig. 18; Plate 3).[31] Despite its apparent strangeness, however, a more leisurely examination of a captured specimen shows that it is quite closely related to *Aysheaia*. It too walks across the seabed on flexible lobopods. In contrast to *Aysheaia*, however, the lobopods of *Hallucigenia* are remarkably long and so slender that the trunk is perched well clear of the sediment. At the end of each lobopod there is a pair of powerful claws. The trunk of *Hallucigenia* is also much narrower than that of *Aysheaia*. Another difference is that in *Hallucigenia* there is a well-defined head that consists of a globe-like mass that hangs downwards. But the really remarkable feature of *Hallucigenia* is the double row of sharp spines that arise from the upper side of the trunk and project up into the overlying water. Each spine is firmly attached to the trunk by the expanded base. *Hallucigenia* is evidently not an animal to be meddled with.

To begin with we have seen only a few solitary specimens of *Hallucigenia*. We then notice the decaying remains of a large, partially dismembered, arthropod lying on the sea floor.[32] It is covered with specimens of *Hallucigenia*, voraciously feeding on this corpse. We count more than thirty already on the carcass, and see several more. *Hallucigenias* busily marching towards it. Like those already feeding, these individuals have been attracted by the smell of decay wafting through the sea water. In a few days, when the arthropod has been picked clean, the *Hallucigenias* will disperse in search of new opportunities for scavenging.

The battery of spines carried by *Hallucigenia* is eloquent testimony to the dangers in the Cambrian seas. But where does the chief menace lie? Looking out the submersible window one of us sees a large shadow stirring in the distant murk, before it slips out of sight. The submersible pursues. In a matter of minutes we are hovering above one of the wonders of the Cambrian world, *Anomalocaris* (Plate 3).[33] Why is it so remarkable? First, it is the largest animal yet encountered, measuring just over a metre in length.

Caught in the glare of the submersible's floodlights, the *Anomalocaris* lies immobile. This allows a net to be carefully lowered over it. Drawn into a special chamber it can now be examined at leisure. The upper surface of the body is dull green with creamy stripes and spots, the lower side pale off-white. Seen from above, the body is divided into two units. The head region bears a pair of large eyes, but is otherwise rather smooth. The trunk is elongate and bears a series of prominent flap-like structures, while at its posterior end there is a spectacular tail-fan. By rotating the animal the investigators can now examine its underside, which yields several surprises. Running along the length of the trunk, on either side of the midline, is a series of legs. Their structure reveals an important clue as to the relationships of *Anomalocaris*. This is because those legs are remarkably similar to the lobopods of *Aysheaia*, soft and flexible yet highly effective for locomotion across the sea floor. It is at the front end of the animal, however, that the real surprises lie. First, there is a pair of prominent appendages (Figs. 11, 21). They show the jointed construction typical of arthropod limbs. Flexible and mobile, these giant appendages are used to grasp and manipulate food. The second surprise is the mouth, which is formed by an extraordinary array of plates (Fig. 20). These encircle the mouth and have sharp prongs that project into the oral cavity. The time travellers take a worm, previously captured, and place it in the chamber with the *Anomalocaris* animal. Within seconds it is grasped by the giant appendages, which guide it to the mouth, where the circle of plates holds on to the struggling prey as it is pushed into the gut to be consumed.

It is now time to release the captured *Anomalocaris* so as to learn more about its behaviour in the natural habitat of the Burgess Shale. The released animal moves swiftly over the soft muds, in search of its prey. It approaches a group of trilobites peacefully grazing on the seabed. With a lunge, one of the trilobites is seized, but although bitten, it manages to struggle free. Before the wounded animal scuttles to safety, we note a large arcuate bite-mark on its right-hand side, near the posterior end (Fig. 41).[34] The next victim is not so lucky. The struggling trilobite is unable to free itself (Plate 3), and as the *Anomalocaris* animal moves away only a few fragments of trilobite exoskeleton litter the sea floor.

The next animal that we encounter is much smaller than *Anomalocaris*, but in some ways it strangely reminiscent. This is *Opabinia* (Fig. 42; Plate 3).[35] At first sight it too looks very peculiar, especially on account of the long nozzle-like structure that snakes ahead of the rest of the animal. As *Opabinia* moves slowly forward, the nozzle is being rapidly moved across the sea floor, stirring up the sediment. The function of this nozzle is revealed when the claw-like apparatus at its end suddenly grasps a writhing worm. Equally promptly the nozzle is folded backwards so that the prey can be stuffed into the mouth located on the underside of the head. On the upper side of the head is another remarkable feature, in the form of five large eyes. The nozzle and five

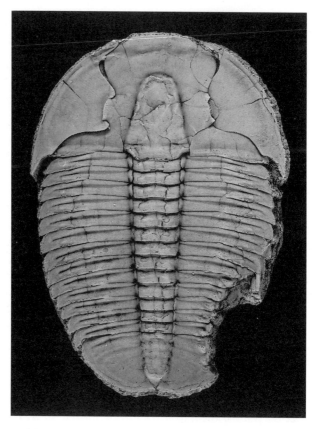

Fig. 41. A Middle Cambrian trilobite with a prominent bite mark on its right-hand side, perhaps caused by an *Anomalocaris* attack. [Photograph courtesy of L. Babcock (Ohio State University, Columbus).]

eyes are certainly peculiar, but as the examination of *Opabinia* continues, we soon realize that this animal must be quite closely related to *Anomalocaris*. Most notable in terms of similarity are the series of flap-like structures arising along the trunk, while at the posterior end there is a tail-fan. The only specimen of *Opabinia* that we find manages to evade capture, but we are almost certain that we can glimpse a view of a series of lobopod-like legs moving beneath the body as it scuttles off.

Of the animals that move across the Burgess Shale sea floor, many are active, especially if they are being pursued or are in pursuit of prey. Others, however, are more sluggish. As the submersible hovers above the seabed we catch our first sight of *Wiwaxia* (Fig. 43; Plate 3).[36] The animal seems to be unaffected by the bright searchlight. Unconcernedly it continues to crawl slowly across the mud, leaving behind a broadly meandering trail. From a distance the most obvious feature of *Wiwaxia* is the double row of elongate

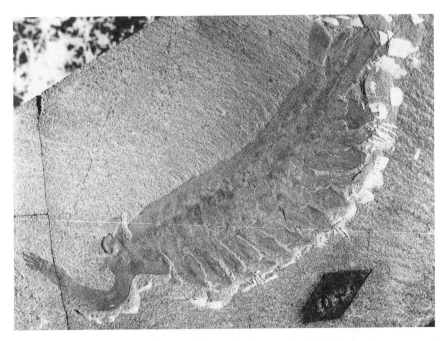

Fig. 42. The Burgess Shale animal *Opabinia regalis*. [Photograph courtesy of H.B. Whittington (University of Cambridge).]

spines that project upwards from the body, in a manner vaguely reminiscent of *Hallucigenia*. At the front and back of the animal the spines are relatively short, whereas in the mid-region they are elongate. The primary role of these spines must be defensive, a conjecture that is supported by the observation that in the specimen one of the spines on the right-hand side has been snapped off, leaving only a short stump.

Wiwaxia, however, is protected by more than its spines. The crew manoeuvre the submersible so that it can scoop up the specimen for a more detailed examination. In the well-illuminated observation chamber *Wiwaxia* is a beautiful golden-yellow colour. This arises from a coat of scale-like structures, known as the sclerites, that cover the entire surface of the animal. Although the primary purpose of the sclerites appears to be defensive, they have one of several distinctive shapes according to which part of the body they cover (see also Fig. 84). On the upper region, between the two rows of spines, the sclerites are arranged in a series of transverse rows. The sides of the animal are covered by another series of sclerites, while along the lower edges of the animal the sclerites arise in a series of bundles, each sclerite having approximately the shape of a banana.

The observation chamber of the submersible is equipped with a remotely controlled set of instruments, broadly similar to the dissecting kit used in a

Fig. 43. The Burgess Shale animal *Wiwaxia corrugata*.

biology class. By careful manipulation one of the scientists is able to detach a sclerite. Once freed, it can be transferred to a microscope stage to be studied in more detail. At the proximal end of the sclerite is a root-like extension that was originally inserted into the body wall. The walls of the sclerite are composed of tough chitin. Under high magnification they reveal a striated microstructure.[37] Somewhat to their surprise, however, the zoologists discover that the sclerite is not solid, but is hollow with a fluid-filled interior.

The *Wiwaxia* animal seems unperturbed by the removal of a sclerite. Indeed, as it grows the smaller sclerites are presumably shed, to be replaced by larger ones that maintain a protective cover. The scientists now guide the

Wiwaxia on to a glass sheet in order to be able to observe its underside. This is revealed to be a broad, soft area, very similar in appearance to the foot of a slug or snail. As in these animals, so in *Wiwaxia* bands of muscular contraction pass along the underside and so propel the animal forwards. As the *Wiwaxia* animal continues to crawl over the glass, its method of feeding is revealed. Periodically the mouth, which is located on the ventral surface but near the front end of the animal, opens. At this point a toothed feeding apparatus (a radula) protrudes, and rasps the underlying surface, so sweeping up any organic material. In general form *Wiwaxia* is rather reminiscent of primitive molluscs, especially in terms of its gliding foot and method of feeding.

Once the *Wiwaxia* animal has been released, the submersible swings round and approaches the Escarpment. Having discovered the infaunal polychaete *Burgessochaeta*, the scientists are now anxious to see if there are any

Fig. 44. The Burgess Shale polychaete *Canadia spinosa*.

epifaunal varieties. Adjacent to some rocks they discover what they are looking for in the form of several specimens of the polychaete *Canadia* (Fig. 44; Plate 4).[38] At first glance the chaetae, which are an iridescent yellow, appear to cover the entire animal. Once a specimen has been captured and then immobilized its overall structure becomes clearer. As in *Burgessochaeta* the body is composed of segments, each of which has soft flexible lobes known as parapodia bearing dorsal bundles of chitinous bristles (the notochaetae), and also ventral bundles (the neurochaetae). In the details, however, *Canadia* differs markedly from *Burgessochaeta*. First, the notochaetae form a spinose array that covers the entire upper surface (see also Fig. 84). The notochaetae arise along transversely elongate parapodia, each bundle overlapping the one behind it, so as to give a tile-like covering to the upper surface of the body. There can be little doubt that this entire arrangement is primarily protective. The neuropodia, in contrast, are much more lobe-like and strongly muscular. Each neuropodium bears a prominent bundle of chaetae, and it is on these structures that the worm walks. Looking at another captured specimen of *Canadia*, the time travellers see that the animal progresses by a series of loco-motory waves passing along the neuropodia. By precise coordination each neuropodium is first placed on the seabed and then pushed back so that the neurochaetae act as levers to push the animal forwards. Finally, at the end of the stroke the neuropodium lifts the chaetae clear of the sediment and swings them forward in preparation for the next shove against the sea floor.

The scientists continue their study of *Canadia*. There are two major sur-prises. First, when a chaeta is tugged free and then placed under the micro-scope it is seen to have a microstructure very similar to that observed in *Wiwaxia*.[39] Evidently there is some evolutionary connection between *Canadia* and *Wiwaxia*. Second, during the dissection of *Canadia* they discover that located between each of the notopodia and neuropodia there is a gill, used for respiration. In itself this is not very surprising, since many polychaetes are equipped with gills. What is surprising, however, is that the gill of *Canadia* is closely similar to the gills of many molluscs (which are known as ctenidia). There seems, therefore, to be also a connection of some sort between *Canadia* and the molluscs. What these various links might be will become clearer in Chapter 7 when we consider new evidence from a new Burgess Shale-type fauna in Greenland, and in particular examine a fossil group known as the halkieriids.

Swimmers and floaters

We could easily spend many more hours searching for new wonders in the Burgess Shale fauna. Time, however, is beginning to run short. The reserves of oxygen are only enough for at most another hour of exploration. Before

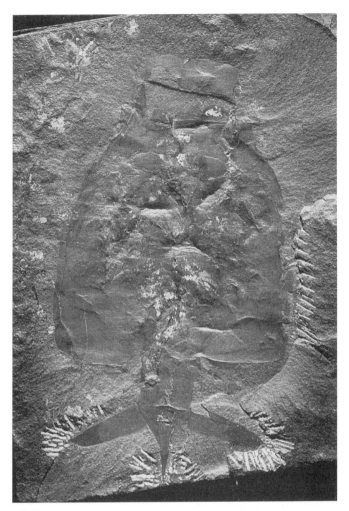

Fig. 45. The Burgess Shale arthropod *Odaraia alata*. [Photograph courtesy of the Smithsonian Institution (Dr D. Erwin).]

returning to the surface the priority is to find those animals that float or swim above the seabed, that is the members of the pelagic ecosystem. The search begins by scanning the waters a few metres above the seabed. Here, too, life is prolific.

First to be encountered is the arthropod *Odaraia* (Fig. 45; Plate 4).[40] By now we have seen such a variety of animals in the Burgess Shale seascape that we are beginning to wonder if there can be any more surprises. Nevertheless, the arthropod *Odaraia* certainly looks very peculiar. In a way somewhat reminiscent of a living lobster, the front of the animal is enclosed in a shield-like structure, known as the carapace. In *Odaraia* the carapace is unusual in that it meets along the midline of the underside, so that the front of the animal is effectively enclosed in a kind of cylinder. From the front end of this carapace there protrude a pair of prominent eyes, each on a stalk. From the rear of the carapace there extend the abdominal segments. Flickering continuously along the lower side are a series of beating appendages. They help to propel *Odaraia* through the water. But the really extraordinary feature of *Odaraia* is the tail. This consists of prominent flukes, vaguely reminiscent of the tail of a whale except that in *Odaraia* there are not two flukes, but three. Moreover, in this arthropod the flukes are not used for actual propulsion, but for balance and steering. In the dimly lit water *Odaraia* is strangely beautiful as with agile swoops it hunts for its prey.

The specimen of *Odaraia* soon moves beyond the searchlight of the submersible. As the vehicle advances slowly forward, we notice a shoal of disc-like animals, hanging almost motionless in the water. As we cautiously approach it is clear that the shoal consists of several hundred individuals. Some are evidently juveniles, others are fully grown adults with diameters of up to 20 centimetres. At first sight the scientists assume, reasonably enough, that they must be some sort of jellyfish. The animals are largely composed of gelatinous tissue, which is semi-transluscent and through which the silvery-white outlines of the internal organs are discernible.

Fig. 46. The Burgess Shale animal *Eldonia ludwigi*.

These animals are known as *Eldonia* (Fig. 46; Plate 4).[41] The underside of the disc is more or less flat, except for a set of feathery tentacular structures that dangle downwards. These are used for feeding on microscopic food particles suspended in the water. Once trapped by the sticky tentacles, the food is transported to the mouth for subsequent digestion. The gut is very distinctive because it forms a prominent coiled organ within the body. It is, moreover, clearly divisible into three sections forming anterior, middle, and posterior units. It is the middle section that is the most obvious because it has conspicuously thick walls. Presumably it acts as a sort of stomach. The other striking feature about the internal structure of *Eldonia* is the set of radial structures, consisting of chord-like strings that are attached to a small ring in the centre of *Eldonia*. They extend outwards, running round the gut, to the margins of the animal. These, however, are not the only radial structures in *Eldonia*. The exterior of the animal, especially towards its edges, is divided into a series of prominent lobes.

Despite first appearances, the entire anatomy of *Eldonia* is obviously far too complex for it to be any sort of jellyfish, which belong to the phylum Cnidaria. Nevertheless, the scientists are quite puzzled by *Eldonia*. In a number of ways it recalls a group known as the sea cucumbers or holothurians. These are relatives of the sea urchins and starfish, all being placed in the phylum Echinodermata. Like most echinoderms, the great majority of sea cucumbers live on or near the seabed. They feed on detritus, using a bush-like mass of tentacles that sweeps across the sea floor. Overall, as the time travellers quickly realize, the arrangement of the sea cucumber tentacles is closely similar to the feeding apparatus of *Eldonia*. A few living sea cucumbers are swimmers and vaguely jellyfish-like. But these animals are otherwise not very similar to *Eldonia* and there is little to support the idea of a close relationship.

To complete the comparison between *Eldonia* and sea cucumbers, one needs to consider in a little more detail their respective internal anatomies. The sea cucumbers have a number of distinctive features, which not surprisingly are also found in the other echinoderms. Most notable is an anatomical arrangement known as the water-vascular system. In essence, this is a complex system of fluid-filled canals, most of which are internal. With certain exceptions, echinoderms have the general peculiarity of having a strongly developed fivefold symmetry. This is perhaps most familiar from the five arms of a starfish. This symmetry is expressed in many parts of an echinoderm. In terms of the water-vascular system it is revealed as five principal branches that converge on a central canal that runs around the gut. Not all of the water-vascular system, however, is internal. An important component protrudes from the branches, through the body wall, to form tentacular extensions on the exterior of the animal. These extensions are known as the tube-feet. On the underside of a starfish, for example, these tube-feet are clearly visible. They have a variety of functions, including walking across rocks and also feeding. What is clear is that whatever resemblance *Eldonia* has to the sea cucumbers, which also have tube-feet, the Cambrian animal has no true water-vascular system. As they watch the shoal drifting through the water, the scientists become increasingly intrigued with *Eldonia*. They come to no final conclusion, but they agree that it must be related in some way to the echinoderms. In the official expedition report they suggest that *Eldonia* may belong to a group that is ancestral to the echinoderms.

As the submersible moves forward, one of the scientists remarks on how surprising it is that we, as vertebrates, are actually quite closely related to the echinoderms. The evidence for this is drawn from several lines of enquiry, notably molecular biology and embryological development. Is there any chance that the common ancestor of chordates (to which phylum we as vertebrates belong) and echinoderms will be discovered in the Burgess Shale fauna? Probably not! By this stage in the Cambrian, evolution has moved on; the Burgess Shale is simply too young for such an animal to be found. The scientists realize already that if they wanted to discover the common ancestor of the chordates and echinoderms they would have to return to the time machine, reset the controls, and travel millions of years further back. Nevertheless, before they departed on their journey there were high hopes that some important new facts might be learnt about the first chordates. They will not be disappointed.

Our attention is caught by a group of lanceolate animals, swimming above the sea floor by flicking their bodies in a series of rapid side-to-side undulations. The submersible follows in rapid pursuit and captures two of the animals, known as *Pikaia* (Fig. 47; Plates 1 and 4).[42] As soon as they can be studied properly in the observation chamber it is clear that they not only are chordates, but must be remarkably primitive. Overall *Pikaia* is yellowish-white in colour, although some of its internal anatomy is just discernible. The specimens are elongate, tapering at each end, and the body is flattened laterally so as to maximize its propulsive force as it swims. Why are the scientists so confident that *Pikaia* is a chordate? There are two main reasons. First, it

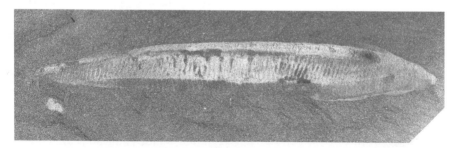

Fig. 47.   The Burgess Shale chordate *Pikaia gracilens*.

Colour plate 1. Both mud-dwellers (infauna) and mud-stickers (sessile epifauna) are shown, with some of the latter in the process of dislodgement by the scoop of the time-travellers submersible. The infauna is dominated by priapulid worms, of which the most abundant was *Ottoia*. In this scene three individuals are visible: one on the floor of the large excavation, another in the process of consuming hyoliths (mid-right), whilst the third is emerging from its burrow and displaying its spinose proboscis (lower right). Two other priapulids are visible in the excavation: the elongate, more-or-less horizontal worm is *Louisella*, shown here in its life position as a sedentary animal occupying an elongate burrow with openings to the overlying sea water at either end. The animal inclined downwards, with its posterior end just emerging from the sea-floor, is an example of *Selkirkia*. It inhabited a parchment-like tube, and in common with other priapulids had a spiny proboscis that was employed, when necessary, for burrowing. The other type of worm visible in the excavation are two examples of the polychaete annelid *Burgessochaeta*, with one individual wriggling on the floor and the other in its burrow with anterior tentacles extending sideways (far left). The sessile epifauna is represented by the enigmatic *Dinomischus* (lower left), the sponge *Vauxia* (blue), the ?cnidarian *Mackenzia* (green), and the ?sponge *Chancelloria* (upper left, purple). Also present is a trilobite (centre) strolling across the sea-floor, and swimming through the water a solitary *Pikaia* (a primitive chordate).

Colour plate 2. This picture emphasizes the epifaunal elements of the Burgess Shale community. Attached to the sea-floor are various types of sponge including the large *Vauxia* (blue; note the lobopodian *Aysheaia* crawling around the edge of the osculum (far mid-right)), *Pirania* (lower left, with prominent projecting spicules), and *Choia* (centre, resembling a pin-cushion). Further in the background are examples of *Dinomischus* (yellow), *Mackenzia* (green), and *Chancelloria* (purple). The prominent group of three frond-like organisms on the mid-upper left are examples of the sea-pen *Thaumaptilon*. Moving across the sea-floor are also two trilobites.

Colour plate 3. In the foreground *Anomalocaris* has captured a hapless trilobite, seized in its anterior giant appendages which are manoeuvring the prey towards the armoured mouth. On the sea-floor from left to centre respectively are a solitary specimen of *Wiwaxia* and three specimens of *Hallucigenia*. Note in both animals the defensive arrays of spines, although the bifid termination in the left individual of *Hallucigenia* is an error. Further to the right is the lobopodian *Aysheaia* with its anterior prongs around the mouth, as well as the primitive arthropod *Opabinia* which is a close relative of the larger *Anomalocaris*. Descending to the sea-floor are two individuals of the arthropod *Marrella*. Also visible in this scene are sessile epifauna in the form of *Dinomischus* (yellow) and the sponge *Vauxia* (blue).

Colour plate 4. The emphasis in this picture is on the swimmers and floaters. In the foreground and ascending upwards is the arthropod *Odaraia*, while higher in the water column are two individuals of the chordate *Pikaia* (left) and the ctenophore *Ctenorhabdotus* (right). The gelatinous discoidal object on the left is *Eldonia*, possibly a primitive echinoderm. On the opposite side is the enigmatic *Nectocaris*. Crawling across the sea-floor is a specimen of the polychaete annelid *Canadia*, whilst the attached forms include the sponge *Pirania* with its elongate spicules upon which are attached some symbiotic brachiopods (which display their marginal setae), and also examples of *Dinomischus*.

propels itself by contracting the muscles of its body wall, which send out a series of waves along the body, in the same manner as a fish. As in many other animals, these muscles are arranged in a repeated longitudinal series; that is, they are segmented. However, in contrast to a phylum such as the annelids, where the muscles of each segment form simple rings, in *Pikaia* and more primitive chordates represented by the fish the arrangement of the musculature is much more complex. The muscles are built up in a sort of cone-in-cone arrangement, to which the term 'myotome' is applied. On the sides of the animal this segmentation is expressed as a series of zig-zags or chevrons. This arrangement of muscles is clearly visible in *Pikaia*. So far as is known the myotomal arrangement of muscles is known only in the chordates. This phylum has another characteristic and unique feature. Running along the upper side of the animal is an internal rod, termed the notochord. The vertebral columns of humans and other vertebrates are based on the notochord, although it is of course considerably elaborated as a bony spine. In very primitive chordates the notochord forms a tough but elastic rod. Why is it present? In brief, it forms what is known as an antagonistic organ, which acts against the contraction of the myotomal musculature. The notochord is clearly visible in *Pikaia*. In comparison with other chordates, however, it shows two unusual features. First, it is located closer to the upper side than is typical. Second, and more importantly, the notochord does not extend all the way to the anterior end, but stops short.

The details of the front end of *Pikaia* are too small to be discerned easily without the help of a microscope. The scientists anaesthetize a specimen and then manoeuvre the animal into position. The head itself consists of a pair of lobate structures, each giving rise to a long slender tentacle. Behind the body, on the first section of the trunk, there arises on either side a series of short appendages. The provisional conclusion is that these appendages are connected to the gills, which in chordates have a very particular arrangement. In this phylum (and in some members of a related group known as the hemichordates) the anterior of the body is perforated by gill slits that connect the anterior gut (which is referred to as the pharynx) to the exterior of the body. To act as gills sea water rich in oxygen is drawn in through the mouth and pumped through the gill-slits, the waste water being expelled. Only careful dissection will confirm that the little appendages of *Pikaia* are linked to the gill-slits, but it seems to be a reasonable idea.

The history of vertebrates is quite well known from the fossil record. Mammals, to which we humans belong, evolved from reptiles about 225 Ma ago, although the appearance of hominids was considerably later (about 4.5 Ma) (Fig. 1). Reptiles can be traced back to the Carboniferous period (about 320 Ma), and must have evolved from an amphibian of some sort. Today this latter group is familiar from animals such as frogs and newts. In the Carboniferous the amphibians looked rather different, some of them reaching a considerable size. Where did the amphibians come from? Largely thanks to wonderfully preserved fossils collected from the Devonian of East Greenland, we now know considerably more about the stages that connect the fish to the amphibians.[43] Before about 370 million years ago, there were no vertebrates on land, and the story of their earlier evolution must be sought in the seas. During Silurian and early Devonian times the marine realm was teeming with many different types of fish. The more advanced types were equipped with jaws, but the more primitive fish were jawless. Some descendants, notably the lamprey, survive to the present day. In the early Palaeozoic, however, there were many other types of jawless fish, many heavily armoured. At some time in either the late Cambrian or early Ordovician, the first fish[44] evolved from an animal not so different from *Pikaia*.

For the scientists, therefore, the recognition of *Pikaia* is momentous. Here is a golden opportunity to investigate some of the earliest stages in the evolution of their own phylum. At first sight the distance from *Pikaia* to a human seems to be almost immeasurable. The differences are indeed major. From an evolutionary viewpoint, however, what matters is that our basic construction as chordates is clearly visible in *Pikaia*. Many of the changes, notably the closure of the gill slits and the development of the four limbs from pre-existing fins, are a direct consequence of the invasion of land in the Devonian. But the possession of the stiffening notochord and muscular myotomes is a necessary first step. The brain of *Pikaia* was very small, but its basic structure would have been similar to that of a primitive fish. Ever since then in the vertebrates there has been, in general, increasing elaboration of this organ, although the increase in brain size of the humans and their immediate ancestors is little short of extraordinary. Nevertheless, in *Pikaia* not only do we see the basic body plan of the chordates, but we can recognize that in its brain there were the first dim stirrings of neural activity that half a billion years later would emerge as fully fledged consciousness.

Catastrophe

While the submersible has been travelling above the sea floor the underlying muds have for the most part been relatively featureless. Here and there, however, we notice fissures running across the seabed. These structures are clear indications that the sea floor is not in a stable condition, but is liable to slumping.[45] Elsewhere, indeed, there are large scoop-like hollows where evidently a piece of seabed has already slumped down-slope into deeper water. What might trigger these episodes of slumping is not easy to determine. The region along which the Cathedral Escarpment runs is, however, known to have been a zone of long-lived tectonic activity.[46] Thus, even minor earthquakes might be sufficient to trigger a movement in these wet sediments.

By chance as the submersible passes over an extensive area of fissures the seabed begins to slump. At first, its motion is hardly perceptible. Soon, however, many square metres of sediment are moving slowly down-slope, away from the escarpment. Some fast-moving arthropods are able to rush to safety, but those animals rooted to the sea floor begin to rock sideways and then topple. As the rate of movement increases it is clear that the sediment is no longer sliding as a single unit but has lost cohesion. It is now a dense cloud of sediment, sweeping rapidly along. At first the submersible remains close to the sea floor, but visibility is becoming rapidly reduced as the flow grows in size. Fearing that the swirling mud may enter and damage the motors, the crew adjust the controls to allow the submersible to rise higher in the water. Now it is possible to see the leading edge of the sediment flow that continues to sweep up animals in its path. We just have time to glimpse a group of *Marrella*s overwhelmed, followed shortly afterwards by several *Pikaia*s that cannot swim fast enough to escape. Behind this leading edge there is a rapidly moving cloud of sediment. Occasionally an arthropod or some other animal is glimpsed at the top of the flow before being re-engulfed by the ebullient turbulent flow.

The distance the cloud of sediment travels is only a few kilometres.[47] As the slope of the sea floor flattens out, so the sediment flow quickly loses its momentum. The sediment, and the thousands of animals carried with it, is rapidly deposited. Although this process of settling will take several days to complete finally, we can already discern that the flow has been transformed into a blanket of sediment. Within it there are entombed thousands of animals. So rapid was their final burial that they have come to rest at all sorts of angles. As the submersible cruises across the seabed there are no signs of movement, no evidence of attempts to escape. Within the sediment the first steps in preservation are under way. Eventually they will produce silvery fossils in a black shale, exposed to the summer sunlight and the eager gaze of Charles Walcott.

Ascent to the surface

Time has almost run out and the submersible begins a slow ascent towards the surface. What new animals shall we encounter on this the last stage of our voyage? The first discovery is a large bag-like animal, with a body made largely of gelatinous tissue and so appearing semi-translucent. This is *Ctenorhabdotus* (Fig. 48; Plate 4).[48] It is quite clearly related to marine animals known today as sea gooseberries or sea combs, which belong to the phylum Ctenophora. *Ctenorhabdotus* is swimming rapidly, and it is easy to see why. Running along the body there is a series of distinct strips, which

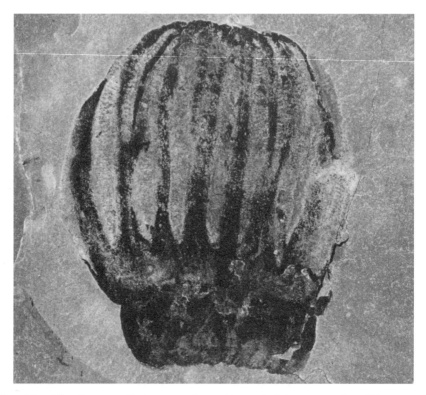

Fig. 48. The Burgess Shale ctenophore *Ctenorhabdotus capulus*. [Photograph courtesy of Dr D. Collins and Royal Ontario Museum, Toronto.]

converge towards one end of the animal where there is a dome-like structure. These strips form structures known as the comb-rows. Along each of them there is a constant flickering movement, set up by the organized beating of many rows of hair-like cilia. Many microscopic organisms, such as *Paramecium*, use the beating of cilia to propel themselves. The cilia, however, are tiny and at first sight it would seem to be impossible for them to drive an animal as big as *Ctenorhabdotus* through the water. The answer lies in the detailed construction of the comb-rows. They are composed of many rows of cilia, each row consisting of a series of cilia that form a paddle-like structure (known as a ctene). By itself the beating of a single ctene would provide rather little propulsive force, but the coordinated beating of several thousand propels *Ctenorhabdotus* rapidly and effectively. What about the dome-like structure? This houses balancing organs that inform the animal of its orientation in the water. If *Ctenorhabdotus* tips, the balancing organs triggers the comb-rows on one side to beat more strongly and the animal rights itself.

Neither of the scientists on board is much of an expert on the ctenophores, but they remember that living ctenophores invariably have eight comb-rows.

In contrast, *Ctenorhabdotus* has 24 such rows. Why *Ctenorhabdotus* had three times the number of comb-rows (and other Cambrian ctenophores had even more) is not certain. Perhaps a reduction to eight rows led to easier coordination via the balancing organ and so improved manoeuvrability. Most living ctenophores feed by trailing long slender tentacles behind the swimming body. When special sticky structures entangle prey, the tentacle is quickly wound back towards the animal and the victim transferred to the mouth. A few ctenophores, however, lack tentacles. They seize their prey using a voluminous mouth. It now seems likely that this type of feeding is actually the more primitive, because *Ctenorhabdotus* has no sign of any tentacles. Its mouth is very wide and muscular. Although we do not see *Ctenorhabdotus* feeding, it seems likely that its prey, when detected, is rapidly pursued and then engulfed by the gaping mouth. Even as the food is seized, the sharp edges of the mouth, which rather surprisingly are made of batteries of fused cilia which act like teeth, are already slicing into the captured animal.

The discovery of a ctenophore in the Burgess Shale is not too surprising. They are generally agreed to be an exceedingly ancient group, which probably originated at about the same time as the cnidarians. Both groups are generally referred to by zoologists as diploblastic. This means that there are two primary (or germ) layers of cells that give rise respectively to an outer ectoderm and the inner endoderm, the latter lining the digestive cavity and also providing the reproductive tissue. Separating these two layers in the cnidarians is a gelatinous mass known as the mesoglea. The mesoglea provides the bulk of the cnidarian body, especially in animals such as the jellyfish, where it is massively developed. Ctenophores are not quite the same. In principle, they too are diploblastic, but the mesoglea also contains muscular fibres. This is significant because it foreshadows the so-called triploblastic arrangement that is found in all animals higher than the cnidarians and ctenophores (as well as the even more primitive sponges). In the triploblasts, the intermediate germ layer forms the so-called mesoderm, which forms a variety of tissues, notably the musculature. Nevertheless, in themselves the ctenophores do not appear to be intermediate between the diploblastic cnidarians and primitive triploblasts such as the flatworm (phylum Platyhelminthes). Rather ctenophores hint at how an intermediate stage of tissue development may have evolved. Unfortunately, the true origins of the ctenophores are still very obscure. To solve this riddle we would almost certainly have to travel back to the Ediacaran seas.

The submersible continues its ascent. It turns out that *Ctenorhabdotus* is by no means the only predator to be found far above the Burgess Shale sea floor. Next to be encountered is *Nectocaris* (Fig. 49; Plate 4).[49] It is very streamlined, and darts through the water with quick flicks of its muscular abdomen. It moves too fast to be easily captured, and we can catch only a few

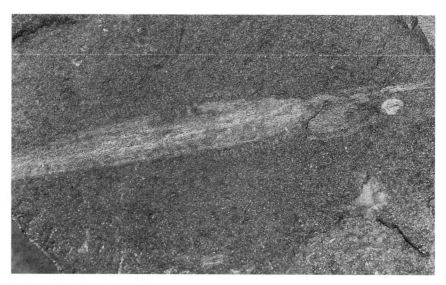

Fig. 49. The Burgess Shale animal *Nectocaris pteryx*.

glimpses. Near the front of the body is a pair of enormous eyes. Just behind them the rest of the head is enclosed in a carapace-like structure, somewhat reminiscent of an arthropod. The abdomen, however, is distinctly unarthropod-like, because running along the upper and lower sides there are prominent fins supported by fin-rays. Such a feature is never seen in the arthropods.

In the modern oceans an important group of predators are the arrow worms (the phylum Chaetognatha). Like *Nectocaris* they dart swiftly through the water. They also have prominent fins supported by fin-rays, but these arise on the sides and posterior end of the body, and in general there is little similarity between the arrow worms and *Nectocaris*. The front end of a chaetognath is little more than a sophisticated grasping machine, consisting of curved rows of dagger-like spines. For many years palaeontologists studying Cambrian sediments have been recovering similar thorn-like microfossils.[50] Now we see the original animal. It is somewhat larger than most living chaetognaths, and bears an impressive array of grasping spines. The specimen we see is consuming a struggling arthropod larva, which has already been forced into its mouth.

The submersible rises to the surface, and with its ballast tanks empty it bobs on the ocean swell. An upper hatch opens and we gulp in the fresh air. It is twilight, with a deep orange glow on the western horizon. Higher in the sky the planet Venus shines, and by turning round we can see the dark silhouette of the island and the outline of the waiting time machine. The submersible turns and we move back towards the island. It is time to go home.

Notes on Chapter 4

1. Precise geochronological dating of the Cambrian using methods of radiometric determination is only in its infancy. In the past few years particular attention has been devoted to the age of the base of the Cambrian. From figures of *c.* 590 Ma, which until recently were widely accepted, the estimate for the beginning of the Cambrian has dropped dramatically, to about 545 Ma. The details can be found in a paper by S.A. Bowring and others (*Science*, Vol. 261, pp. 1293–8 [1993]), in which precise determinations of U/Pb (uranium/lead) ratios in zircons are presented. A number of earlier papers were in fact already strongly pointing to such an age for the base of the Cambrian (e.g. work by Compston and others reported in the *Journal of the Geological Society, London*, Vol. 149, pp. 171–184 [1992]). This emphasis on a much younger date for the Lower Cambrian has led to proposals that the Cambrian was much shorter in duration than was once thought. Curiously, rather less attention has been given to the date of the top of the Cambrian (and by definition the beginning of the succeeding Ordovician period), but a paper by J.L. Bonjour and others in *Chemical Geology (Isotope Geoscience Section)* (Vol. 72, pp. 329–36 [1988]) indicates that this date is also younger than widely believed; so the total length of the Cambrian may be in the region of 70 Ma. There are, as yet, only limited radiometric data for dates within the Cambrian, but a figure of *c.* 525–520 Ma for the Burgess Shale seems to be reasonable.

2. Our understanding of the distribution of the Cambrian sediments around Laurentia owes much to the American palaeontologist A.R. (Pete) Palmer. He identified an inner detrital belt, adjacent to land, and consisting mostly of sands and silts, often with glauconite, a mineral formed during or soon after the deposition of the sediments. Further offshore there is the median carbonate belt, which is rich in limestones and other carbonates that show evidence for deposition in generally fairly shallow water. Beyond this zone is the outer detrital belt, which evidently faced the open ocean. In this area mostly muds and silts accumulated. There is abundant evidence for submarine slopes and even cliffs where material from shallower water sometimes cascaded or slumped downslope into deeper water. It is in the outer detrital belt that many of the Burgess Shale-type faunas are located, although others evidently occurred in much shallower water. An overview of the distribution of these faunas is given in my papers in *Philosophical Transactions of the Royal Society of London* B (Vol. 311, pp. 49–65 [1985]), and in *Transactions of the Royal Society of Edinburgh: Earth Sciences* (Vol. 80, pp. 271–83 [1989]).

3. The idea of islands along the rim of the Cathedral Escarpment was mentioned by I.A. McIlreath (*Special Publications of the Society of Economic Paleontologists and Mineralogists (SEPM)*, Vol. 25, pp. 113–24 [1977]) on the basis of evidence of sediment by-passing over the shallow-water shelf. In my paper on the Burgess Shale community (*Palaeontology*, Vol. 29, pp. 423–67 [1986]) I speculate (p. 427) that the fauna may have lived beneath this island, at the toe of the Cathedral Escarpment.

4. The exact depth at which the Burgess Shale fauna lived is only approximately known. The presence of algae suggests a position well within the photic zone (the region into which sunlight penetrates), and thus a depth of less than about 100 m. Some of the most conspicuous algae occur in slabs of shale without other fauna, and it is possible that these algae (especially *Marpolia spissa*) derive from shallower depths than the rest of the fauna. Using a local biostratigraphy based on trilobites, W.H. Fritz (*Proceedings of the North American Paleontological Convention*, Part I, pp. 1155–70 [1971]) presented evidence for a depth of deposition a little before the Burgess Shale of *c.* 200 m. Allowing, therefore, for sedimentary infill a figure of *c.* 100 m for the fauna seems possible.

5. After a period of relative neglect, several research groups have renewed their interest in the priapulids. A very active German team, led by Volker Storch, has published extensively. Recent key papers include those in *Journal of Morphology* (Vol. 220, pp. 281–93 [1994]) and *Invertebrate Biology* (Vol. 114, pp. 64–72 [1995]). There is also an important overview, published in Russian (but with an English summary), by V.V. Malakhov and A.V. Adrianov as *Cephalorhyncha—a new phylum of the animal kingdom* (KMK Scientific Press, Moscow, 1995).

6. A description of *Ottoia* and the other Burgess Shale priapulids was published by me in *Special Papers in Palaeontology* (Vol. 20, pp. i–iv, 1–95 [1977]).

7. This is almost complete speculation because the transport of the Burgess Shale fauna destroyed all the traces made by the animals, including burrow systems. Some living priapulids live in U-shaped burrows. In terms of the fossil record, S. Jensen (*Lethaia*, Vol. 23, pp. 29–42 [1990]) has suggested that Lower Cambrian burrows from Sweden were constructed by priapulids. Unpublished work by Soren Jensen has documented burrows made by living priapulids, with the recognition of features that may also be identifiable in the fossil record.

8. Most workers suggest that hyoliths are closest to the Mollusca. The Burgess Shale hyoliths were redescribed by E.L. Yochelson (*Journal of Paleontology*, Vol. 35, pp. 152–61 [1961]). Some comments were also made subsequently by B. Runnegar in *Lethaia* (Vol. 13, pp. 21–5 [1981]).

9. See note 6.

10. Although these stud-like structures were recognized as long ago as 1977, only recently has it been realized that they derive from the skin of priapulid worms. A paper I published (*Zoological Journal of the Linnean Society*, Vol. 119, pp. 69–82 [1997]) reviews the evidence for this interpretation, and also provides a description from electron micrographs of a remarkably preserved specimen of this group, generally referred to as palaeoscoledians.

11. See note 6.

12. The polychaetes of the Burgess Shale were described by me in *Philosophical Transactions of the Royal Society of London* B (Vol. 285, pp. 227–274 [1979]).

13. Information on this rather extraordinary discovery by J. Vacelet and N. Boury-Esnault can be found in *Nature* (Vol. 373, pp. 333–5 [1995]).

14. The sponges of the Burgess Shale were described by J.K. Rigby (*Palaeontographica Canadiana*, Vol. 2, pp. 1–105 [1986]).

15. This association between sponge and brachiopods has never received detailed study. It is commented on by H.B. Whittington (*Proceedings of the Geologists' Association*, Vol. 91, pp. 127–48 [1980]), and also illustrated in *Atlas of the Burgess Shale* (ed. S. Conway Morris) (Palaeontological Association, London, 1982).

16. A comprehensive redescription of the Burgess Shale chancelloriids is unfortunately not yet available, although they are being studied afresh by S. Bengtson and D. Collins, using in part spectacular new specimens discovered by the Royal Ontario Museum, Toronto.

17. S. Bengtson has been instrumental in arguing that this method of spicule construction debars the chancelloriids from being true sponges. He has also drawn attention to the strong similarities, especially in mode of construction, to the variety of Cambrian spicule-like fossils that belong to groups such as the halkieriids, sachitids, and siphogonuchitids. He places all of them in a group, perhaps roughly equivalent to a phylum, known as the Coeloscleritophora. A review of our present understanding of the coeloscleritophorans can be found in the major monograph by S. Bengtson and others published in *Memoirs of the Association of Australasian Palaeontologists* (Vol. 9, pp. 1–364 [1990]). Some doubt has subsequently been thrown on the coherence of this group. In particular N.J. Butterfield and C.J. Nicholas (*Journal of Paleontology*, Vol. 70, pp. 893–9 [1996]) have reinterpreted the chancelloriids in terms of a sponge model.

18. A description of *Thaumaptilon* (and also *Mackenzia*) is given in my paper in *Palaeontology* (Vol. 36, pp. 593–635 [1993]).

19. See note 18 for Chapter 2.

20. A full systematic description of the Ediacaran fronds is not yet available, but a useful description of *Charniodiscus* was published by R.J.F. Jenkins and J.G. Gehling in *Records of the South Australian Museum (Adelaide)* (Vol. 17, pp. 347–59 [1978]).

21. These ideas of a basal anchoring disc giving way to an ability to retract into a burrow have not been tested. The former idea might be amenable to investigation in the setting of a flume-tank. Cambrian trace fossils consisting of a vertical burrow and rich in the calcareous spicules derived from the breakdown of the pennatulacean tissue might also provide evidence in support of this suggestion.

22. See note 18.

23. The description of *Dinomischus* was published by me in *Palaeontology* (Vol. 20, pp. 833–45 [1977]). The affinities of *Dinomischus* still remain unresolved. The Polish palaeontologist J. Dzik has suggested that it may be related to the Burgess Shale animal known as *Eldonia* and another Cambrian fossil *Velumbrella* (see his paper in *The early evolution of Metazoa and the significance of problematic taxa* (ed. A.M. Simonetta and S. Conway Morris), pp. 47–56 (Cambridge University Press, 1991). This idea, however, remains speculative. *Dinomischus* has also been reported from the Burgess Shale-type biota of Lower Cambrian age in Chengjiang, China by J-Y. Chen and others (*Acta Palaeontologica Sinica* (Vol. 28, pp. 58–71 [1989]); see note 23 of Chapter 5.

24. The detailed description of *Marrella* was published by H.B. Whittington in *Bulletin of the Geological Survey of Canada* (Vol. 209, pp. 1–24 [1971]).

25. Two detailed papers on *Olenoides* have been published by H.B. Whittington. The first was in *Fossils and Strata* (Vol. 4, pp. 97–136 [1975]); the second, which dealt with topics such as moulting and locomotion, appeared in *Palaeontology* (Vol. 23, pp. 171–204 [1980]).

26. The connection between various types of Palaeozoic trace fossil and the activities of trilobites moving across or within the sediments of the sea floor has been rather contentious. For the most part, palaeontologists accept that the trackways known as *Diplichnites* were made in some circumstances by trilobites, although other arthropods almost certainly contributed to the record of this trace fossil. Much more dispute, however, has surrounded another trace fossil, known as *Cruziana*, which although widely attributed to the activity of trilobites, may have been made by other types of animal. R. Goldring in *Geological Magazine* (Vol. 122, pp. 65–72 [1985]) discusses how *Cruziana* may have been made.

27. A detailed description of *Naraoia* was published by H.B. Whittington in *Philosophical Transactions of the Royal Society of London* B (Vol. 280, pp. 409–43 [1977]). One of the first species of soft-bodied animal to be described from the Chengjiang deposits in South China, of Lower Cambrian age, were examples of *Naraoia*. The initial description by W-T. Zhang and X-G. Hou in *Acta Palaeontologica Sinica* (Vol. 24, pp. 591–5 [1985]) has not yet been complemented by a detailed description, but some additional remarks by L. Ramsköld and others can be found in *Lethaia* (Vol. 29, pp. 15–20 [1996]). In this latter paper the authors argue that supposed segmental divisions of the carapace are artefacts that have arisen during the processes of compaction. Examples of *Naraoia* are also known from other Middle Cambrian deposits; there are, for example, reports from Utah by me and R.A. Robison (*The University of Kansas Paleontological Contributions*, Paper 122, pp. 23–48 [1988]).

28. The description of *Aysheaia* by H.B. Whittington is to be found in *Philosophical Transactions of the Royal Society of London* B (Vol. 284, pp. 165–79 [1978]). It is now realized that *Aysheaia* is only one of a wide variety of so-called lobopodians, which are regarded as primitive arthropods. The other main example in

the Burgess Shale is *Hallucigenia*, which also occurs in the Chengjiang deposits with a number of related species (see X-G. Hou and J. Bergström in *Zoological Journal of the Linnean Society*, Vol. 114, pp. 3–19 [1995]).

29. The proposal that *Aysheaia* may have fed upon sponges comes from H.B. Whittington (see note 28), where he noted an intriguing association between this species of lobopodian and various sponges. It is possible, however, that the association is not original. As a result of the transport of the fauna from the pre-slide environment, where the animals lived, there has been a mingling of individuals that in life were apart. J. Monge-Najera in a paper in *Zoological Journal of the Linnean Society* (Vol. 114, pp. 21–60 [1995]) points out that sponge spicules have not been identified in the gut of *Aysheaia*.

30. A summary of current thought in this area was presented by G.E. Budd in *Lethaia* (Vol. 29, pp. 1–14 [1996]); see also Chapter 7.

31. The history of our understanding of *Hallucigenia* was reviewed in Chapter 3, where the relevant references are also given.

32. The evidence for such congregation of *Hallucigenia* on decaying material was published by me in *Palaeontology* (Vol. 20, pp. 623–40 [1977]).

33. The story of how our knowledge of *Anomalocaris* has improved, including the recognition of legs, is told in Chapter 3.

34. The evidence for bite-marks in Cambrian trilobites and the remarkable fact that there is a preponderance of them on the right-hand side of the carapace is presented by L.E. Babcock in *Journal of Paleontology* (Vol. 67, pp. 217–29 [1993]). Another paper relevant to this discussion is by D.M. Rudkin in *Royal Ontario Museum Life Sciences Occasional Paper* (No. 32, pp. 1–8 [1979]), where he describes possible evidence for predation in trilobites from the *Ogygopsis* Shale, which is located a few kilometres from the Burgess Shale. Rudkin makes the prescient suggestion that *Anomalocaris* may have been responsible for the wounds observed, although it was not for some years that the jaw of *Anomalocaris* was correctly identified, so giving added credence to this proposal. H.B. Whittington and D.E.G. Briggs (*Philosophical Transactions of the Royal Society of London* B, Vol. 309, pp. 569–609 [1985]) discuss the possible feeding action of *Anomalocaris*.

35. *Opabinia* caught the eye of the distinguished biologist G.E. Hutchinson and he discussed its possible relationships within the arthropods in a paper in the *Proceedings of the United States National Museum* (Vol. 78 (article 11), pp. 1–24 [1930]). A detailed reassessment awaited, however, the work of H.B. Whittington, who published an account in *Philosophical Transactions of the Royal Society of London* B (Vol. 284, pp. 165–97 [1978]). More recently G.E. Budd (*Lethaia*, Vol. 29, pp. 1–14 [1996]) has presented evidence for hitherto overlooked legs in *Opabinia*.

36. I published a redescription of *Wiwaxia* in *Philosophical Transactions of the Royal Society of London* B (Vol. 307, pp. 507–86 [1985]). Since then *Wiwaxia* has been discovered in a number of other Cambrian deposits, including those in Guizhou, South China (see Y-L. Zhao and others, *Acta Palaeontologica Sinica*, Vol. 33, pp. 359–66 [1994]) and also north-west Canada (see N.J. Butterfield, *Nature*, Vol. 369, pp. 477–9 [1994]). The phylogenetic position of *Wiwaxia* and its relationship to the Lower Cambrian halkieriids is discussed at some length in Chapter 7.

37. This microstructure was first identified by N.J. Butterfield and described in *Paleobiology* (Vol. 16, pp. 287–303) [1990]). He drew attention to its striking similarity to the microstructure in the chaetae of polychaetes, and thereby concluded that *Wiwaxia* belonged to this group of the Annelida. As I explain in Chapter 7, this interpretation needs some important qualifications, but it appears to be correct that *Wiwaxia* has an important role in understanding how the annelid body plan evolved from a mollusc-like predecessor.

38. The Burgess Shale polychaetes, including *Canadia*, were described in a paper I published in *Philosophical Transactions of the Royal Society of London* B (Vol. 285, pp. 227–74 [1979]).

39. See note 37. In this paper N.J. Butterfield also comments on the microstructure of the chaetae in *Canadia*.

40. The detailed description of *Odaraia* is by D.E.G. Briggs and may be found in *Philosophical Transactions of the Royal Society, London* B (Vol. 291, pp. 541–84 [1981]).

41. The systematic redescription of *Eldonia* is in an unpublished Ph.D. thesis (D. Friend. *Palaeobiology of Palaeozoic medusiform stem group echinoderms.* University of Cambridge, 1995). This work represents a major advance on our earlier understanding, although the paper by J.W. Durham in *Journal of Paleontology* (Vol. 48, pp. 750–5 [1974]) contains useful information.

42. No detailed redescription of *Pikaia* is yet published. At present I am working with D.H. Collins on a major monograph on this animal.

43. Details of this work are given in note 1 of Chapter 1.

44. Reports of fish scales by G.C. Young and others from the Upper Cambrian of Australia (*Nature*, Vol. 383, pp. 810–12 [1996]) are, therefore, very significant. A hitherto unrecognized diversity of fish in the Ordovician, as for example reported by I.J. Sansom and co-workers in *Nature* (Vol. 379, pp. 628–30 [1996]), also supports the view that fish evolved in the Cambrian.

45. The evidence for slumping of regions of the sea floor upon which the Burgess Shale fauna was living was first articulated by H.B. Whittington in his introductory paper to the Cambridge Campaign, published in the session on Extraordinary Fossils in *Symposium of the North American Paleontological Convention (1969), Chicago,* (Part I, pp. 1170–1201 [1971]), and was supported by the sedimentological investigations of D.J.W. Piper (*Lethaia*, Vol. 5, pp. 169–75 [1972]). The varied orientation of the specimens, some of them end-up, and the presence of graded bedding are both strong indications of transport and loss of cohesion of the sediment as it travelled down-slope before being deposited in the so-called post-slide environment.

46. A review of this zone, known as the Kicking Horse Rim, is given by one of the leaders of the Burgess Shale excavations in 1966 and 1967, J.D. Aitken in *Bulletin of Canadian Petroleum Geology* (Vol. 19, pp. 557–69 [1971]).

47. The calculation that the Burgess Shale was transported only a short distance is based on some rather circuitous reasoning which I gave in a paper in *Palaeontology* (Vol. 29, pp. 423–67 [1986]).

48. A description of *Ctenorhabdotus* and two other species of Burgess Shale ctenophore was published by me and D.H. Collins in *Philosophical Transactions of the Royal Society of London* B (Vol. 351, pp. 279–308 [1996]). In this paper we noted that there were unpublished reports of fossil ctenophores from the Chengjiang deposit in South China. Since then a specimen has been illustrated in the book by J.-Y. Chen and others entitled *The Chengjiang Biota: a unique window of the Cambrian explosion* (National Museum of Natural Science, Taiwan, 1996). The text is in Chinese, but the photographs give an excellent introduction to this extraordinary biota.

49. Only a single specimen was known when I described this animal in *Neues Jahrbuch für Geologie und Paläontologie* (Monatshefte 12, pp. 705–13 [1976]). D.H. Collins of the Royal Ontario Museum, Toronto, has shown me a number of specimens collected by his expeditions. He believes that these specimens may also represent either *Nectocaris* or a similar animal.

50. The likely affinity of the thorn-like protoconodonts to the chaetognaths has been articulated by the Polish palaeontologist H. Szaniawski in *Journal of Paleontology* (Vol. 56, pp. 806–10 [1982]), and *Fossils and Strata* (Vol. 15, pp. 21–7 [1983]). D.H. Collins has shown me new material from his collection that is strongly reminiscent of the chaetognaths; there is also an undescribed specimen that was found by Walcott and is now in the Smithsonian Institution (USNM 199540).

CHAPTER 5

The search for new Burgess Shales

Greenland

It is the morning of 4 July, 1989. Slightly to our surprise we are all standing safely on the ground. Beside us is almost a tonne of expedition equipment, which we have just unloaded from an aeroplane. A few minutes earlier we were still wheeling in the sky, anxiously looking for an area safe enough to put the aeroplane down. This is far from easy because we are in the virgin wilderness of North Greenland. In this remote area there are very few places where an aeroplane can land. Most of the tundra is simply too bumpy, too rocky, or just waterlogged. Several times our pilot Bjarki has just touched down, the wheels of the aeroplane bouncing against the ground. In a split second Bjarki has decided that the risk is too high, the twin engines roar loudly, and we climb steeply back into the air and safety. Finally, having banked steeply, Bjarki makes a decision and with consummate skill he brings the aeroplane in to land on what seems to be an impossibly short stretch of ground. As we step down on to the tundra we can see the huge inlet from the Arctic Ocean, a few kilometres away, known as J.P. Koch Fjord and named after one of the intrepid early explorers of Greenland. The surface of the Fjord shimmers white; it is always covered with sea ice. Although in the summer it is flecked with melt-water pools the colour of ethereal turquoise, the ice itself never breaks up, let alone completely melts. Beyond the Fjord, further west, are mountains capped with glaciers and snow.

Why have we travelled to the high Arctic, to Peary Land situated at 83° North? The story (Fig. 13) really starts in 1984, when the Geological Survey of Greenland was in the middle of preparing the first detailed geological map of North Greenland. The professionalism of the Survey belies the magnitude of the task. In a few short summer seasons an area as large as England was covered. This is seemingly an impossible task, but the Arctic offers some advantages. High above the Arctic Circle at this time of year the continuous daylight, although not necessarily continuous sunlight, makes for long working days. In addition, there is very little vegetation and the geology is laid out with extraordinary clarity. The Survey expedition operates from a base camp, but most of the geologists move from fly camp to fly camp by helicopter, sometimes covering hundreds of square kilometres in a few weeks. In an initial foray to J.P. Koch Fjord, some outcrops of shale exposed in hills above the Fjord were examined by two British geologists, Tony Higgins and

Jack Soper, working for the Survey. Although neither was a palaeontologist, a sharp lookout for fossils was always kept because they can often provide a precise age for the rocks. This is essential if the geological history is to be unravelled in its correct order. In the shale outcrops the geologists found fossils, but they were not ones they could easily identify. Every evening, however, there is radio contact between the survey teams, not only for safety, but also to report on the latest progress. Talking on the radio that evening to John Peel, a palaeontologist working in another area, Jack Soper described the fossils as being 'like a bunch of grass'. The true importance of these fossils became clear only back in Copenhagen, Denmark, where the Survey is based. The specimens turned out to be well-preserved sponges, and their discovery was a hint that even more remarkable fossils might be unearthed.

The mapping programme continued, and the following year the region was revisited, again by Tony Higgins and a student assistant, Neil Davis. Stopping near the sponge locality, but this time on the other side of the hill, Tony Higgins found abundant fossils. There was little time for collecting, but a handful of slabs was quite sufficient to show that they had stumbled on a fossil bonanza.

The next year, that is in 1986, John Peel brought some of the Peary Land fossils to a special conference in Uppsala, Sweden, convened to discuss Cambrian life. He was already highly excited by this discovery, because apart from the trilobites there were clearly a number of soft-bodied fossils, similar to those of the Burgess Shale.[1] When John Peel explained to me the general geological setting of this new discovery, which is now known as the Sirius Passet fauna, it was immediately apparent that there were some strong parallels with the Burgess Shale. In Cambrian times, North America and Greenland formed a continent (known as Laurentia) which lay across the palaeo-equator and was isolated from the other Cambrian continents by seas and oceans (Fig. 50). Immediately to the south of the Greenland locality there is an extensive area of limestones and other carbonate sediments. Back in the Cambrian this was an area of extensive and shallow seas. Incidentally, in Cambrian times these tropical seas would have lain to the west of Sirius Passet: not only has Laurentia migrated by continental drift, but it has also rotated by about 90°. The Sirius Passet fauna, however, was deposited in muds that lay in deeper waters. The junction between the carbonates and the muds, which now form a highly fissile shale, is very sharp. Although the carbonates do not appear to have formed an escarpment as visibly dramatic as that in the Burgess Shale, there is nevertheless evidence for a submarine cliff or slope of some sort. Large chunks from the edge of the carbonate platform occasionally fell or slid into the adjacent basin, where the Sirius Passet fauna lived.[2] Recall that the Burgess Shale fauna also flourished in muds beside the carbonates of the towering Cathedral Escarpment.

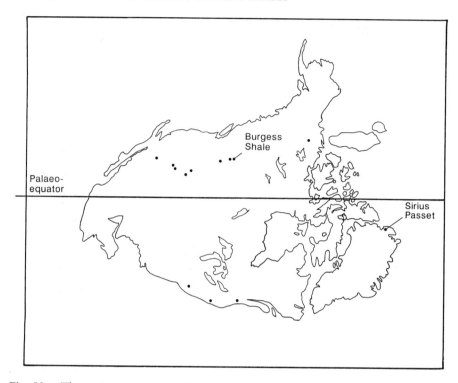

Fig. 50. The ancient continent of Laurentia. Each dot represents a fauna of Burgess Shale type, with the Burgess Shale itself and Sirius Passet specifically identified. The other localities include those in California, Nevada, Utah, Idaho, British Columbia, Northwest Territories, Vermont, Pennsylvania, and Tennessee. Note the position of the palaeo-equator and the broadly concentric distribution of these faunas round the craton. The outline of Laurentia, as it might have been seen by a hypothetical Cambrian satellite, would have only a vague correspondence to the map shown here, which is based on modern geography. All the Burgess Shale-type faunas are marine, and so the coastline lay inward of these sites. In addition, the Cambrian is marked by a steady transgression of the sea so that over time the shoreline retreated and the area of dry land became correspondingly reduced in extent. [Reproduced with permission from fig. 2 in S. Conway Morris and J.S. Peel (1995). Articulated halkieriids from the Lower Cambrian of North Greenland and their role in early protostome evolution. *Philosophical Transactions of the Royal Society of London* B, Vol. 347, pp. 305–58.]

On the strength of the information available we decided that it was imperative to organize an expedition to Peary Land. The first trip, in 1989, was fraught with uncertainties and imponderables. The locality with the Sirius Passet fauna was perched above J.P. Koch Fjord. Running eastwards to another fjord known as Brainard Sund is a huge valley, then nicknamed Muddy Valley, but now termed Sirius Passet. We knew that at Brainard Sund there was a usable landing strip, but that was more than 25 km from where we wanted to be. If we had to make our base camp there, most of our time during the short Arctic summer would be spent trudging up and down Sirius

Fig. 51. North Greenland, 4 July 1989. The plane prepares to leave, having flown us to J.P. Koch Fjord.

Passet, moving in food and fuel and carrying out any fossils that we managed to collect. After our nerve-racking search for a landing site we were thus doubly relieved when our pilot, Bjarki, finally managed to put us down substantially closer to the fossil locality (Fig. 51). Even so, our problems were by no means finished. Not only did we have to cross a large river and find a suitable campsite closer to the fossils, but we also needed to build a proper landing strip.

We were naturally very anxious to visit the locality at the first opportunity. Would it be as rich as the Burgess Shale, or would there only be a few small outcrops and a handful of fossils? Even before we established our final camp, we walked across country for an initial reconnaissance. Most of the way was across open, windswept tundra, but as we climbed into the hills some slopes were still deep in snow. Once at the locality (Fig. 52) it was clear that we were on the verge of a major discovery. Before us was a large hill slope. Near its crest were projecting outcrops of rock, but most of the hill was covered by tens of thousands of thin slabs of dark shale. Looking first at those slabs scattered along the foot of the hill we could see at once that there was an abundance of fossils. Within minutes of our arrival, one of the team, Paul Smith, bent down to pick up a slab and then asked me, 'Simon, is this of any interest?' Here indeed was an extraordinary fossil, a flattened, slug-like creature with a covering of scales. My immediate hunch was that this must be an articulated specimen of a halkieriid (Fig. 53), a group that hitherto had been known only from its scattered sclerites. Here, all the sclerites were beautifully

Fig. 52.   The Sirius Passet locality. It is early in the season, and so there is plenty of snow. The figure in the foreground is John Peel.

articulated to form a coat like chain mail, which quite clearly could be seen to form a series of distinct zones across the body (Figs. 67, 84).[3] Even more extraordinary, however, was that at either end of the body there was a large shell. More puzzling still was that the one at the posterior end looked uncannily like a brachiopod. This is a group of shellfish that we briefly encountered earlier in the discussion of the mud-stickers in Chapter 4, with examples perched amongst the spicules of the sponge *Pirania* on the Burgess Shale sea floor. At that time, of course, we had no real inkling of the scientific importance of the halkieriids and their relative from the Burgess Shale, *Wiwaxia*. I shall return to the significance of this discovery in Chapter 7.

Collecting continued, with occasional new specimens of halkieriids turning up, amongst an abundance of other forms. It soon became apparent that as in the Burgess Shale the arthropods were the most abundant component of the fauna. Rather curiously, there is only one species of trilobite, whereas in the Burgess Shale there are about twelve. Of the arthropods without calcified exoskeletons, some are quite similar to Burgess Shale species, but overall the similarities are not very marked. Many of the arthropods are large, reaching at least 50 centimetres in length.[4] The appendages are occasionally well preserved, and in some species various remains of the internal anatomy are evident. In addition, there are various worms. These include polychaete annelids and large priapulids. One very striking fact is that there are very few taxa with shelly skeletons: only one species of trilobite, rare hyoliths, a number of sponges with prominent spicules, a few small brachiopods, and no

Fig. 53. One of the first specimens to be discovered at the Sirius Passet locality, the halkieriid *Halkieria evangelista*.

echinoderms or molluscs. In contrast, in the Burgess Shale although the abundance of shelly fossils in terms of total numbers is quite low, their variety is quite considerable. Why the difference between the two faunas? Any suggestions are tentative, but some evidence suggests that the number of shelly species drops dramatically in areas of the sea floor with very low levels of oxygen. Perhaps this is the clue to the differences between the Sirius Passet and Burgess Shale faunas?

During the 1989 expedition, the weather was generally poor. Snow, rain, fog, and cold winds all limited the amount of time we could spend collecting. The time when we hoped to be picked up had come and gone, but the weather continued to be bad: low cloud and fog banks prevented any useful excursions. We knew that the aeroplane was in the region, putting out supplies for the famous Sirius Sledge Patrol, the crack corps of the Danish Armed Forces, which uses dog-pulled sledges for epic training exercises across the frozen winter wastes of North and East Greenland. Sheltering in our tents because the weather was so poor, we had expected the aeroplane to return to the base camp at Station Nord. We could hardly believe our ears when first we heard a distant drone, and then out of the low cloud and fog the aeroplane emerged unannounced. Radio contact was established. The aeroplane zoomed low over our camp, and then turned to make the first landing on our new airstrip. Within two hours the camp was packed, the boxes full of fossils loaded, and we were lurching back down the strip. Rising above a foggy and cold J.P. Koch Fjord we headed eastwards, on the first leg of our journey home.

The first trip to Sirius Passet in 1989 demonstrated that a huge cache of fossils remained to be collected. Since then we have returned twice, first in 1991 (again in a group of four) and most recently in 1994 (only John Peel and myself). On both occasions the weather proved much better, especially in 1991 when there were days of unbroken sunshine. The 1991 expedition led to the largest of the collections being made, and about 4000 fossils were transported back to Copenhagen. The 1994 trip was also successful in this respect, but its principal purpose was to explore the surrounding country to see if there were additional soft-bodied fossil localities. There were some intriguing reports, based on earlier reconnaissances by the Geological Survey of Greenland, of possible occurrences on the far side of J.P. Koch Fjord. In addition, standing above the Sirius Passet locality and looking eastwards towards Brainard Sund it is easy to trace the sharp contact between the pale-coloured carbonates of the platform and dark shales of the adjacent basin. Perhaps there were additional localities, teeming with new soft-bodied fossils? Only a helicopter is capable of easily reaching these areas. The flight was spectacular, but however promising the rocks looked from the air as we circled the outcrops, once on the ground the results were always disappointing. So far as we know the Sirius Passet fauna remains unique.

From our main base camp and its nearby airstrip the locality is only about an hour's walk into the hills, where scree-covered slopes retain snow and ice in all but the warmest summers. Once at the fossil site the view is superb: the eye travels along J.P. Koch Fjord to its entrance with the Arctic Ocean, clearly visible a hundred kilometres away (Fig. 54). The chore of quarrying and excavation itself is largely obviated, because over the centuries a mantle of fallen slabs has covered the hillside as scree. The rock breaks into thin sheets, many of which bear fossils. On the slopes, all one has to do is scan the slabs, rejecting those of no interest and placing them into neat stacks. As the search progressed, so the hillside began to resemble a strange shrine, with small piles of shale dotted all over the surface. Perhaps one in thirty slabs has a fossil worth keeping, and maybe only ten or so fossils collected on any one day are of major importance, but as the days pass the collection grows into thousands of specimens. Now only the best or rarest specimens are kept.

Fig. 54. J.P. Koch Fjord, North Greenland, looking north towards the Arctic Ocean.

How does one protect such delicate material on its fragile, brittle slabs? Collecting becomes a slick operation (Fig. 55). First the slab is trimmed of excess shale. Then cardboard boxes are unfolded and foam pads cut out from large rolls. Layer by layer the foam and the precious fossils alternate, until the box is finally filled, numbered, and carried back to camp.

Palaeontologists are fortunate among scientists to travel to remote areas. There is no doubt that the chief delight of the subject is to explore the novelty of vanished worlds. But in addition to see wilderness and experience a remoteness that, however intangible, transports one far beyond the humdrum affairs of ordinary life. The high Arctic is a strange paradox of expected harshness and unexpected fertility. In the evanescent summer, flowers dot the landscape. Bright purple and yellow stonecaps are complemented by Arctic buttercups and poppies, the last forming luminous patches of incandescent yellow with an other-worldly transparency as the low sun shines through the flowers shaking in the wind. On the tundra, trees are of course, absent, but the Arctic willow forms strange twisting patterns as its narrow branches crawl over the ground. Animals are rare: some birds such as skuas and the snowy owl, the occasional musk-ox and snow-hare; while footprints in the snow reveal not only fox but also wolves. But such a description may give too strong an impression of fecundity. The final impression is one of harshness and uncompromising severity, of emptiness. The sun may be low in the sky, but it seems to bleach the landscape. Here it looks like a remote and distant star. The clarity of air brings distant, eroded cliffs into sharp focus, the terrane can look almost lunar. It is an alien world, and one gets a glimpse

Fig. 55. Sirius Passet locality, John Peel trimming specimens prior to packing them.

of what might be the fate of a planet circling a cooling sun, where life lingers on the edges of a frozen world, before slipping to annihilation and extinction as the star turns to embers.

Laurentia

Earlier in this chapter I explained how similar the settings of the Burgess Shale and Sirius Passet faunas are: both flourished on muddy seabeds in fairly deep water. These locations looked out towards the open ocean, but were also adjacent to the carbonate platforms that encircled the ancient continent of Laurentia. In this sense these two faunas are by no means unique. In fact Burgess Shale-type faunas are known from many areas of the world and span a considerable part of the Cambrian period (Fig. 56).[5] The majority of these faunas are known from North America. Although in part this is because of the abundant exposures of suitable rocks, it must also be due in part to the relative ease of accessibility, together with the scientific traditions and

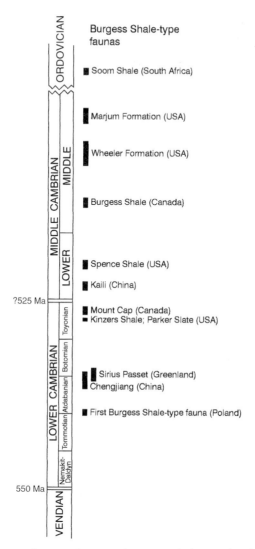

Fig. 56. Distribution of some of principal Burgess Shale-type localities in the Lower and Middle Cambrian.

number of palaeontologists scouring the ground. At present about 27 other Burgess Shale-type faunas are known from the Lower and Middle Cambrian of Laurentia. If one plots their position (Fig. 50), then it is clear that they more or less encircle the continent. Important gaps, such as those in north-west Canada and North Greenland have recently been filled in, and we can certainly expect more discoveries. Not every gap, however, will be filled: in many parts of North America the Cambrian sediments are deeply buried and beyond easy observation, even by deep boreholes. Despite the concentric positioning of these Burgess Shale-type faunas, their locations are not always quite the same as they were in the Cambrian. The Burgess Shale, for example, has actually been transported eastwards as part of a huge slice of rock, by a process known as thrusting, during mountain-building.

Nevertheless, the general concentric pattern is clear. Detailed investigations of the geology also show that many of these faunas are in settings similar to those of the Burgess Shale and Sirius Passet. Unfortunately, not all the faunas are still accessible for collecting. For example, two quite famous localities are those of the Kinzers Shale[6] in Pennsylvania, and Parker Slate[7] in Vermont, both of Lower Cambrian age. Not many soft-bodied fossils are known from these sites, but some of them are very intriguing. Regrettably the quarries that yielded these fossils have long since been closed or filled in, and the only known specimens are from museum collections. New finds continue to trickle in from some of the localities in Utah and Idaho,[8] and such rare discoveries are largely thanks to the perseverance of dedicated amateur collectors, perhaps most notably the Gunther family, who generously are willing to share their finds with professional palaeontologists. Any locality with Burgess Shale-like preservation qualifies, but at some sites the total number of such fossils recovered is so far very small. In 1993 we visited one such locality in southern California.[9] There was considerable excitement, therefore, when one of my colleagues discovered a fossil worm. Our delight turned more to surprise when subsequently we realized that this was the opposite side (that is the counterpart of a specimen) that had been collected more than ten years earlier.

The generally concentric distribution of the Burgess Shale-type faunas around the ancient continent of Laurentia means that in principle there should be a high predictability of where to find new localities. Nevertheless, there are complications. In many places the Cambrian strata are either covered by younger rocks or have been entirely eroded away. Even where they are exposed they may be altered by heat and pressure into metamorphic rocks. Adjacent to both the Burgess Shale and Sirius Passet the sediments become strongly affected by metamorphism, which has turned them into cleaved slates in which any soft-bodied fossils are very unlikely to survive.

What then are the prospects for new discoveries? Several years ago, Nick Butterfield was sent some shale samples, which were thought to be Precambrian and thus similar in age to sediments he had been studying as a postgraduate at Harvard University. The actual samples were from boreholes drilled in an area to the east of Norman Wells in the Northwest Territories of Arctic Canada. It was apparent almost immediately that the shales must be Cambrian, because they contained diagnostic shelly fossils. Nick Butterfield then decided to use his expertise in disaggregating sediments, which had led to such interesting results in the case of *Wiwaxia* (see Chapter 7), on the samples. The results were remarkable.[10] The fossils included sclerites of *Wiwaxia*, more beautifully preserved than those from the Burgess Shale

Fig. 57. Isolated sclerite of *Wiwaxia* from the Mount Cap Formation, north-west Canada. [Photograph courtesy of N.J. Butterfield (University of Cambridge).]

(Fig. 57). Quite clearly the sclerites are hollow and thus comparable to the sclerites of the halkieriids. Even more astonishing finds were fragments of arthropod limbs, so well preserved that at first sight they might be confused with preparations from living specimens (Fig. 58). It is not yet clear what fragments of limb belong to which arthropods, but the exquisite structure of these appendages with their delicate hair-like processes indicates that they derive from crustacean-like animals and were capable of highly sophisticated filtering of the sea water for suspended particles of food.

Nick Butterfield also discovered that the sedimentary units that yielded these extraordinary fossils are not restricted to the subsurface, but are exposed on the surface, to the west of Norman Wells in the Mackenzie Mountains. There seemed to be a real chance of discovering yet another Burgess Shale-like fauna. In 1994 he and a colleague embarked on an expedition to this remote area of Canada, using a helicopter to reach inaccessible localities. Unfortunately, nothing as dramatic as the faunas of the Burgess

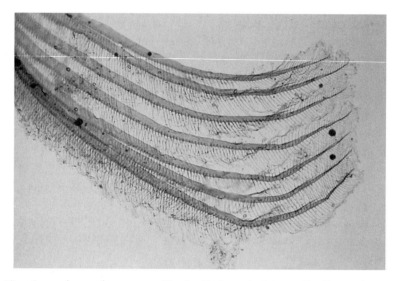

Fig. 58. Appendages of crustacean-like fossils from the Mount Cap Formation, north-west Canada. [Photograph courtesy of N.J. Butterfield (University of Cambridge).]

Shale or Sirius Passet was unearthed.[11] Nevertheless, this region is by no means exhausted of potential, and further surprises may well emerge from the Mackenzie Mountains.

Chengjiang

Although the majority of Burgess Shale-type faunas are from Laurentia, they are not restricted to this Cambrian continent. In particular, there is one locality that certainly rivals the Burgess Shale and Sirius Passet assemblages. This is the famous Chengjiang fauna, best known from hillside exposures at Maotianshan about 50 kilometres south-east of Kunming, Yunnan province, China. Curiously, the first soft-bodied fossils from this region were described in 1912, that is a year after Charles Walcott's initial publications on the Burgess Shale. They were found by a French geologist, Henri Mansuy (Fig. 13),[12] but for some peculiar reason his initial contributions have hardly ever been acknowledged. Further finds were reported in 1957 by Pan Kiang,[13] but the real impetus began in 1984 when, one day before the Sirius Passet fauna was discovered in Greenland, the Chinese palaeontologist Hou Xianguang, from Nanjing, stumbled on the principal locality at Maotianshan. Since then there have been extensive excavations, with labour hired from adjacent villages. Several separate research teams are now studying the Chengjiang fauna.[14]

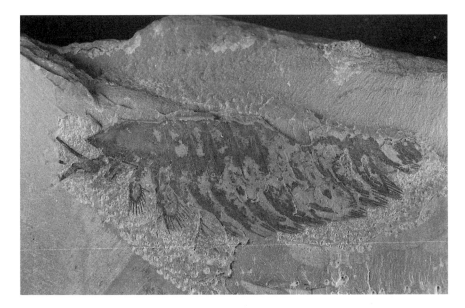

Fig. 59. The Chengjiang arthropod *Leanchoilia illecebrosa* from South China. [Photograph courtesy of Hou Xianguang (Museum of Natural History, Stockholm and Institute of Palaeontology, Nanjing).]

In appearance many of the Chengjiang fossils are spectacular. They are preserved as reddish-brown impressions on a yellow shale, which in its unweathered state is a dark grey colour. What is particularly interesting is the rather strong faunal similarities to the Burgess Shale fauna. This is despite the fact that Chengjiang is not only of Lower Cambrian age and thus somewhat older (Fig. 56), but it was deposited on a continent (the South China craton) that was probably situated several thousand kilometres from Laurentia. There is, as might be expected, an abundance of arthropods (Fig. 59), some of which have very well-preserved appendages.[15] There are also examples of *Anomalocaris*[16] and *Hallucigenia*.[17] A rather remarkable feature of the Chengjiang fauna is an abundance of other lobopodian animals, relatives of *Hallucigenia*. It was the discovery of one of these Chengjiang fossils that gave the first clear indication that my earlier reconstruction of *Hallucigenia* must be upside down (Fig. 19). One of the Chengjiang lobopodians, known as *Microdictyon* (Fig. 60)[18] is especially noteworthy. For a number of years Cambrian palaeontologists had been recovering tiny phosphatic fossils, with a characteristic net-like form (Fig. 61).[19] Their zoological relationships were a complete mystery. What nobody would have predicted is that each phosphatic disc formed a sort of 'shoulder-pad', one above each lobopod of the *Microdictyon* animal. Apart from these relatively unfamiliar arthropods, many of which are remarkably primitive and so are throwing light on the first

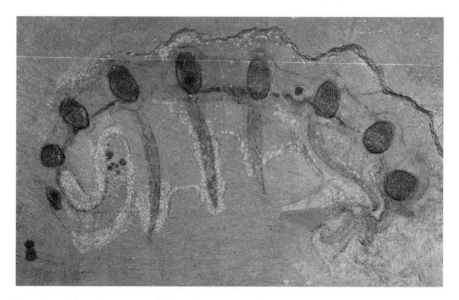

Fig. 60. The Chengjiang lobopodian *Microdictyon sinicum* from South China. [Photograph courtesy of Hou Xianguang (Museum of Natural History, Stockholm and Institute of Palaeontology, Nanjing).]

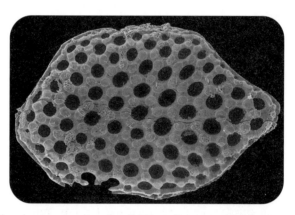

Fig. 61. The phosphatic microfossil *Microdictyon*, which is now recognized as an integral part of a lobopodian (see Fig. 60). [Photograph courtesy of S. Bengtson (Museum of Natural History, Stockholm).]

steps of arthropodization, there are also superbly preserved trilobites, some of which have their appendages visible.[20]

The Chengjiang worms include a variety of priapulids,[21] but they tend to be noticeably smaller than those of the Burgess Shale. So far it appears that no polychaete annelids have been discovered, although they are known from the Sirius Passet fauna, which is probably only slightly younger (Fig. 56).[22]

Neither does Chengjiang appear to contain any sclerite-bearing metazoans similar to either the halkieriids or *Wiwaxia*, although it is surely likely that they will be discovered. The enigmatic *Dinomischus* (Fig. 35) does occur in Chengjiang, but the specimens are conspicuously larger than those from the Burgess Shale.[23] Another similarity between these two faunas is the presence of the medusoid-like *Eldonia*.[24] This animal is particularly abundant in the Chengjiang assemblages. Perhaps the most exciting of recent discoveries is an animal, as yet only known from a single specimen, that appears to be a close relative of the Burgess Shale chordate *Pikaia*. The Chengjiang fossil (Fig. 62), known as *Cathaymyrus* (literally 'Chinese eel'),[25] is of particular interest because at the anterior end the gill slits are clearly visible. In contrast, in *Pikaia* such structures have not been seen, possibly because the Burgess Shale has experienced significantly more sedimentary compaction that obliterated the gill slits. As noted above, however, it is likely that the short appendages just behind the head in *Pikaia* are connected to the gill slits. The discovery of

Fig. 62. The Chengjiang fossil *Cathaymyrus diadexus*, interpreted as the oldest chordate yet identified. The figure was electronically prepared to combine the images of part and counterpart of the specimen (courtesy of Dudley Simons).

Cathaymyrus may also be important for throwing some light on a very strange fossil from Chengjiang, known as *Yunnanozoon*.[26] When this animal was given its first detailed description it was hailed as a chordate,[27] but more recent studies have revealed serious problems with this analysis. It is possible that *Yunnanozoon* occupies a more primitive position within a group known as the hemichordates.[28]

How is it that the faunas of Chengjiang and Burgess Shale are relatively similar despite their separation from each other in time, by perhaps 15 Ma, and space in the form of a major ocean that was many thousands of kilometres wide? In both cases the answer might be tied in some way to the depth of the sea. Some years ago it was realized by American palaeontologists, such as Mike Taylor, that trilobites collected from deeper-water deposits of Upper Cambrian age in areas such as Nevada (Fig. 63) were remarkably similar to trilobites preserved in similar environments on the edge of the Chinese continent.[29] This similarity, however, did not extend to those trilobites living in

Fig. 63. The Hot Creek Range in Nevada, USA. The rocks in the foreground are shales and yield Upper Cambrian trilobites that lived in deep water and are very similar to those found in China. The rocks on the skyline are limestones and yield very different types of trilobite that lived in much shallower water.

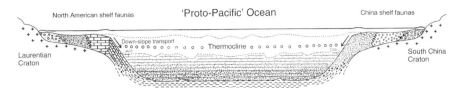

North American shelf faunas   'Proto-Pacific' Ocean   China shelf faunas

Down-slope transport   o o o o o o o o o Thermocline o o o o o o o o o o o o

Laurentian
Craton

South China
Craton

Fig. 64. Hypothetical cross-section of the Cambrian ocean that separated Laurentia from South China. The ocean was thermally stratified, with a thermocline separating the warm surface waters from the deeper, much cooler bottom waters. Deep-water trilobites are found as fossils in sediments that accumulated on the edges of both continents; in Cambrian times they could migrate across the ocean floor. In the shallow water of each continent trilobites adapted to warm, sunlit seas flourished. They could not cross the temperature barrier of the thermocline, but when dead their remains could be transported into deeper water by sediment slumping.

the shallower waters on each continent; they had little in common. What seems to be the best explanation is that the trilobites living in the deeper waters were adapted to cool, dark conditions. They were unable to migrate into the warm, sunlit environments of the shallow shelf seas that rimmed each continent. There was, however, nothing to prevent these deep-water trilobites migrating across the deep ocean floor that separated South China from Laurentia (Fig. 64). This suggests that such deep-water faunas originally occupied huge areas of sea floor. Only the remnants of this trilobite distribution survive, tacked on to the edges of the continents, because nearly all this sea floor was subsequently destroyed. The principal mode of destruction is by the processes of subduction, whereby the ocean floor of one tectonic plate is forced beneath another, such as is occurring today between part of the Pacific Ocean floor and South America. The subduction zone therefore dips into the Earth's interior. The subducted oceanic crust is ultimately carried several hundred kilometres deep into the mantle before it is finally broken up and incorporated into the surrounding material.

Perhaps a somewhat similar explanation applies to the Chengjiang and Burgess Shale faunas? Very early in the Cambrian period the predecessors of these faunas probably flourished in shallow water,[30] where food supplies were most abundant. As the Cambrian progressed, however, levels of competition continued to rise, so that the less successful species were displaced into deeper water where competition was less ferocious. Such displacement, however, opened the possibility of migration across the ocean floor to other areas. It is also thought that the rates of evolutionary change may be slower in deeper-water environments. This might help to explain some of the similarities between the Chengjiang and Burgess Shale faunas, despite the fact that the latter is considerably younger (Fig. 56). Indeed, in some ways the Burgess Shale fauna might be viewed as more of a relict, an archaic survivor of evolutionary events that mostly took place earlier in the Cambrian.[31]

What other localities are there in South China? The Chengjiang site at Maotianshan appears to be the richest and most productive, but comparable occurrences are known elsewhere in Yunnan province, notably at Meishucun[32] and Haikou.[33] In addition, there have more recently been exciting developments at a locality known as Kaili, situated in Guizhou province. This is a Middle Cambrian fauna, but its scientific description is still at a relatively early stage.[34] It is inevitable that new discoveries will be made. Vast areas of South China remain to be searched properly. There are already hints of future discoveries in Siberia, especially in the north and north-east of this huge region. The realization that the Burgess Shale-type faunas are distributed concentrically around Laurentia may also provide a powerful focus for new searches. Fresh finds and new fossils will lead to rejuvenation of ideas and interpretations. Nevertheless, taken together the Burgess Shale-type faunas have a very distinctive character. All are dominated by arthropods, although for the most part trilobites are unimportant. The faunas are also rich in worms, especially priapulids, and generally there is a diverse assemblage of sponges. Quite often there are sclerite-bearing metazoans like *Wiwaxia*, and medusoid-like creatures such as *Eldonia*. There will be plenty of surprises in the next few years, but we are surely already in a position to consider the general importance of these faunas.

Notes on Chapter 5

1. We announced the first discovery of the Sirius Passet fauna in *Nature* (Vol. 326, pp. 181–3 [1987]).
2. Some preliminary comments on the setting of the Sirius Passet fauna can be found in a paper by J.S. Peel and others published in *Rapport Grønlands Geologiske Undersøgelse* (Vol. 155, pp. 48–50 [1992]).
3. The preliminary description of this articulated halkieriid by J.S. Peel and me appeared in *Nature* (Vol. 345, pp. 802–5 [1990]).
4. The descriptions of the arthropods by G.E. Budd are now being published, and to date include papers in *Nature* (Vol. 364, pp. 709–11 [1993]) and *Transactions of the Royal Society of Edinburgh: Earth Sciences* (Vol. 86, pp. 1–12 [1995]).
5. I reviewed the distribution of these faunas some years ago in *Transactions of the Royal Society of Edinburgh: Earth Sciences* (Vol. 80, pp. 271–83 [1989]). Some of the more recent discoveries are reviewed below.
6. The fauna from the Kinzers Shale has never received a systematic redescription, but significant contributions may be found in works by L.D. Campbell (*Journal of Paleontology*, Vol. 45, pp. 437–40 [1971]), D.E.G. Briggs (*Journal of Paleontology*, Vol. 52, pp. 132–40 [1978]), and J.K. Rigby (*Journal of Paleontology*, Vol. 61, pp. 451–61 [1987]).
7. The Parker Slate also lacks an overall synthesis. One of its most interesting fossils, *Emmonsaspis*, had been widely interpreted as a chordate. My re-examination, published in *Palaeontology*, (Vol. 36, pp. 593–635 [1993]), suggested that this fossil was more likely to be related to the frond-like animals that flourished in the Ediacaran assemblages.
8. A very useful summary of the various finds in Utah is given by R.A. Robison in *The early evolution of Metazoa and the significance of problematic taxa* (ed. A.M. Simonetta and S. Conway Morris), pp. 77–93 (Cambridge University Press, 1991).

9. This was an outcrop of the Lower Cambrian Latham Shale in Providence Mountains. Our guide was Mary Droser, a geologist in University of California, Riverside, who showed considerable panache in getting our field vehicle up some daunting desert tracks.

10. The paper by N.J. Butterfield was published in *Nature* (Vol. 369, pp. 477–9 [1994]).

11. The results were nevertheless far from insignificant; the paper by N.J. Butterfield and C.J. Nicholas may be found in *Journal of Paleontology* (Vol. 70, pp. 893–9 [1996]).

12. The paper on the palaeontology of Yunnan by H. Mansuy was published in *Mémoires du Service Géologique de l'Indo-Chine* (Vol. 1 (Part 2), pp. 1–146 [1912]). The Burgess Shale-like fossil is discussed on p. 31, and illustrated on Plate 4, fig. 6. Henri Mansuy has an interesting history, and I am very grateful to Philippe Janvier of the Museum National d'Histoire Naturelle, Paris for giving me a full account of this individual and so freely sharing his information. Mansuy's background was unpretentious. He was a mason's son, and on moving to Paris from the Meuse he was at first a policeman and subsequently a tailor. He became interested in the natural sciences, but also a follower of the anarchist thinker Peter Kropotkin. These latter sympathies seem to have created considerable difficulties, but eventually Mansuy came to the notice of Raoul Verneau, a professor of anthropology in Paris, who found Mansuy a place in the Geological Survey of Indochina in 1901, where he became an official geologist in 1904. In 1909 another geologist, J. Deprat, arrived in Hanoi. The two men became friends, but Deprat's career was later ruined when in 1917 he was accused of fraud, specifically the salting of collections with anomalous trilobites. Why Deprat should have been so foolish is obscure, and Philippe Janvier wonders if a falling-out between Deprat and Mansuy led the latter to malicious action. An analysis of this 'Affaire Deprat' has been published by M. Durand-Delga (in *Travaux du Comité Français d'Histoire de la Géologie* 4: 117–212 [1990]). Deprat himself wrote a novel (*Les chiens aboient*), under the pen-name of Herbert Wild. In his account of the scandal the names of the principal characters are in pseudonymous code.

   Mansuy's work in Yunnan was carried out with the head of the Indochina Survey, Honoré Lautenois. The first fruits of this work on the Cambrian faunas near Chengjiang were published in 1907 (*Annales des Mines* (March–April 1907), pp. 1–209), and it is likely that the first soft-bodied fossils were found during these expeditions, which thus predated even Walcott's collections on Mount Stephen.

13. The paper by K. Pan was published in *Acta Palaeontologica Sinica* (Vol. 5, pp. 523–6 [1957]).

14. There is now quite an extensive literature on the Chengjiang biota. Useful overviews are available in several publications. These include two chapters in *The early evolution of Metazoa and the significance of problematic taxa* (ed. A.M. Simonetta and S. Conway Morris) (Cambridge University Press, 1991) by J-Y. Chen and B-D. Erdtmann (pp. 57–76), and X-G. Hou and J. Bergström (pp. 179–87) respectively. Also useful are the papers by X-G. Hou and others in *Zoologica Scripta* (Vol. 20, pp. 395–411 [1991]), D. Shu and L. Chen (*Journal of Southeast Asian Earth Sciences* (Vol. 9, pp. 289–99 [1994]), and J-Y. Chen and others in *National Geographic Research and Exploration* (Vol. 7, pp. 8–19 [1991]). Most recently is an outstanding colour atlas of this biota published by J-Y. Chen and others entitled *The Chengjiang biota: a unique window of the Cambrian explosion* (National Museum of Natural Science, Taiwan, 1996). The Chinese text has no English summary.

15. Key papers in this respect are by J-Y. Chen and others (*Science*, Vol. 268, pp. 1339–43 [1995]), X-G. Hou and others (*Acta Palaeontologica Sinica*, Vol. 28, pp. 42–57 [1989]), and D. Shu and others (*Alcheringa*, Vol. 19, pp. 333–42 [1995]).

16. The discovery of walking appendages in *Anomalocaris* is documented by X-G. Hou and others in *GFF* (Vol. 117, pp. 163–83 [1995]). Another relevant paper is by J-Y. Chen and others in *Science* (Vol. 264, pp. 1304–8 [1994]).

17. A description of the Chengjiang *Hallucigenia* can be found in the paper by X-G. Hou and J. Bergström (*Zoological Journal of the Linnean Society*, Vol. 114, pp. 3–19 [1995]), which discusses a number of the Cambrian lobopodians.

18. An exhaustive description of *Microdictyon* can be found in the paper by J-Y. Chen and others (*Bulletin of the National Museum of Natural Sciences, Taiwan*, Vol. 5, pp. 1–93 [1995]).

19. The problem of understanding *Microdictyon* before the discovery of the associated soft parts was succinctly discussed by S. Bengtson and others in *Problematic fossil taxa* (ed. A. Hoffman and M.H. Nitecki), pp. 97–115 (Cambridge University Press, 1986).

20. The appendages of the redlichiacean trilobites, a characteristic component of Lower Cambrian faunas in China, are described by D. Shu and others in *Beringeria Special Issue* (Vol. 2 (Morocco '95), pp. 203–41 [1995]).

21. A useful update on the priapulids is given by X-G. Hou and J. Bergström in *Lethaia* (Vol. 27, pp. 11–17 [1994]).

22. The precise date of both the Chengjiang and Sirius Passet faunas is not yet resolved. In general Chinese workers have emphasized an Atdabanian age, even though shelly faunas from some distance beneath the Chengjiang faunas are clearly Upper Atdabanian (see the monograph by Y. Qian and S. Bengtson in *Fossils and Strata* (Vol. 24, pp. 1–156 [1989]), which would suggest a Botomian age is more likely. In the case of the Greenland faunas dating is frustrated by the scarcity of stratigraphically useful fossils. There is only one taxon of trilobite, but its presence has been used to infer a late Atdabanian age by A.R. Palmer and L.N. Repina (*University of Kansas Paleontological Contributions* (Vol. 3, pp. 1–35 [1993]).

23. The available reconstruction of *Dinomischus* differs radically from the Burgess Shale material because arising upwards from the goblet-like body is an enormously elongate tube that the Chinese palaeontologists interpret as an anal chimney, albeit of extraordinary length. Examination of many specimens of *Dinomischus* from Chengjiang by the kindness of several workers, including J-Y. Chen, D. Shu, and X-G. Hou, has not persuaded me that this interpretation is correct. In my opinion the supposed anal chimney is simply the distal part of the stalk folded beneath the body so as to project upwards. Interestingly, the Chinese specimens of *Dinomischus* tend to be somewhat larger than those from the Burgess Shale, but unfortunately the material I have examined does not appear to reveal new details that could help to resolve the systematic affinities of this animal.

24. A description of *Eldonia* is given by J-Y. Chen and others in *Acta Palaeontologica Polonica* (Vol. 40, pp. 213–44 [1995]).

25. The paper by D. Shu and others was published in *Nature* (Vol. 384, pp. 157–58 [1996]).

26. When it was first described, the affinities of *Yunnanozoon* remained unresolved (see the paper by X-G. Hou and others in *Zoologica Scripta*, Vol. 20, pp. 395–411 [1991]. With the discovery of new specimens J-Y. Chen and others proposed a place not only within the chordates, but in a position significantly more derived than *Pikaia* (see *Nature*, Vol. 377, pp. 720–2 [1995]). One of Chen's co-authors, J. Dzik, also published his own account of the chordate interpretation of *Yunnanozoon* in *Acta Palaeontologica Polonica* (Vol. 40, pp. 341–60 [1995]).

27. In the 'News and Views' article that accompanied the reinterpretation of *Yunnanozoon* S.J. Gould wrote a hyperbolic review, and declared that it is 'a beautifully preserved and unambiguously identified chordate' (see *Nature*, Vol. 377, pp. 681–2 [1995]).

28. The reinterpretation by D. Shu and others of *Yunnanozoon* as a hemichordate (*Nature*, Vol. 380, pp. 428–30 [1996]) seems to be more consistent with the evidence. I am not convinced, however, that it is the correct interpretation, although I entirely agree with the refutation of a chordate affinity. Shu Degan generously allowed me to review his Chengjiang collection of *Yunnanozoon* in Xi'an, and on the basis of these observations I would question whether the anterior structure can be resolved as a proboscis comparable to the hemichordate acorn worms. There is also the unresolved problem as to why the putative gill slits are so widely spaced in this animal, whereas in the acorn worms and cephalochordates (including *Cathaymyrus*) they are much more closely spaced.

29. These ideas were reviewed in several papers, of which the one by M.E. Taylor and R.M. Forester (*Bulletin of the Geological Society of America*, Vol. 90, pp. 405–13 [1979]) is perhaps the most accessible. Some further implications of this model of Cambrian palaeo-oceanography were explored by A.W.A. Rushton and me in a *Geological Society of London Special Publication* (No. 38, pp. 93–109 [1988]).

30. The notion of shallow waters being the cradle of evolutionary novelty and diversification has received considerable attention in recent years with the documentation of on-shore–off-shore trends during the Phanerozoic. A useful review of these concepts is given by J.J. Sepkoski and P.M. Sheehan in the book *Biotic interactions in Recent and fossil benthic communities* (ed. M.J.S. Tevesz and P.L. McCall, pp. 674–717 (Plenum, New York, 1983). In the specific context of Cambrian faunas the relevant paper is by J.F. Mount and P.W. Signor in *Geology* (Vol. 13, pp. 730–32 [1985]).

31. The evidence for evolutionary conservatism in the Burgess Shale-type faunas, with the paradoxical observation that although they are our best evidence for the dramatic nature of the Cambrian 'explosion' they are trapped in a sort of deep-water 'museum' and thus are more like echoes of that event, is reviewed in my paper in *Transactions of the Royal Society of Edinburgh: Earth Sciences* (Vol. 80, pp. 271–83 [1989]).

32. The Meishucun occurrence is documented by X-G. Hou and W-G. Sun in *Acta Palaeontologica Sinica* (Vol. 27, pp. 1–12 [1988]).

33. The discovery of a Chengjiang fauna at Haikou is reported by H.-L. Luo and others in *Acta Geologica Sinica* (Vol. 71, pp. 97–104 [1997]).

34. An important set of papers dealing with the Kaili fauna and reporting some of the results of Y-L. Zhao and others can be found in Part 3 (pp. 263–375) of Volume 33 [1994] of *Acta Palaeontologica Sinica*.

CHAPTER 6

The significance of the Burgess Shale

Wonderful life?

By now it will be clear that the Burgess Shale is of the greatest significance in understanding a particular aspect of the history of life: that is, the nature of the explosive diversification of animal life during the Cambrian period. So, for palaeontologists, especially those who are trying to understand the Cambrian, the Burgess Shale fauna is of the first importance. But is it only of interest to the expert and the specialist? Can we in fact claim that the Burgess Shale has an even wider importance? As we saw in the opening chapter, the acceptance of the fact of evolution has led neither to a consensus on mechanisms nor to an agreement as to its implications. At first sight it would seem to be distinctly surprising that the Burgess Shale itself should throw new light on these problems and in doing so should necessitate a reorientation of views, a re-emphasis of priorities, and a reformulation of the debate. Nevertheless, according to some scientists our new knowledge of the Burgess Shale necessitates not only a profound reappraisal of the way we view the processes of evolution, but also their consequences. If this opinion did turn out to be correct, then the Burgess Shale fauna would indeed be of pivotal and special importance in our understanding of evolution. It would remain, of course, as an exemplar, but one of such importance that it could act as the touchstone for a new view of evolution. This, in effect, is the main purpose of the book (*Wonderful life*)[1] written by Steve Gould. In support of the notion that the Burgess Shale fauna should play a major role in such a debate three main lines of argument are apparent:

(1) First, the work in Cambridge by Harry Whittington and his team revealed what appeared to be a remarkable range of animal design in the Burgess Shale. At first sight it seemed to be very difficult to accommodate a significant number of these bizarre-looking fossils in known groups. The diversity of forms certainly reinforced the evidence for the magnitude of the 'Cambrian explosion'. Steve Gould accepted this at face value and proposed that what appeared to be an incredible range of animal types might require us to think of some novel type of evolutionary mechanism.

(2) Steve Gould proposed that the sheer range of animal types (this range of anatomy and morphology is referred to as disparity) was at its maximum

during the Cambrian. He went on to argue that the disparity thereafter declined towards the present day. Paradoxically, what appears to us today to be an amazing variety of animals is, in Gould's opinion, an impoverished remnant of the former glories of the past.

(3) It is always fun to imagine the 'What if?' of history. Suppose that Christianity had become the state religion of China? At one time this looked quite possible. What then? Imagine that the Confederate forces had won the battle at Gettysburg in the American Civil War. In fact, they almost did. What would have been the consequences for the history of the United States? Suppose that Napoleon had died as a child in Corsica. Would not the history of Europe have been far less hideous and traumatic if this evil man had never survived? But is this anything more than an intellectual game? In *Wonderful life* Steve Gould asked his readers to imagine what might be the outcome of rerunning the Cambrian explosion. Would the world today look much the same? Would there be the same end products? You and me, mice and horses, whales and eagles?

As is the case with much of Gould's popular writing, the arguments are presented with some verve and flair. But in the case of *Wonderful life* there seem to be serious doubts as to whether any of these three main themes will stand up to critical scrutiny. In part this is because even in the few years since *Wonderful life* was published we have learnt considerably more. Thus if, as seems to be the case, we can begin to document the origin of body plans in the fossil record and recognize that such evolutionary steps do not involve macroevolutionary jumps, so it is necessary to question Gould's proposals concerning new mechanisms of evolution, perhaps restricted to the Cambrian interval, and hitherto overlooked by a scientific community dominated by neo-Darwinians. The debate initiated by Gould on disparity has similarly led to a fruitful series of interactions, and it may be too early to draw the final conclusions. Nevertheless, the evidence to date does not support his metaphor of an 'inverted cone of life' reflecting a dramatic decline of disparity since the Cambrian. But at the heart of *Wonderful life* are Gould's deliberations on the roles of contingencies in evolution. Rather than denying their operation—and that would be futile—it is more important to decide whether a myriad of possible evolutionary pathways, all dogged by the twists and turns of historical circumstances, will end up with wildly different alternative worlds. In fact the constraints we see on evolution suggest that underlying the apparent riot of forms there is an interesting predictability. This suggests that the role of contingency in individual history has little bearing on the likelihood of the emergence of a particular biological property.

The Burgess Shale undoubtedly reveals much that hitherto had been unsuspected concerning the richness of marine life in the Cambrian. Certainly the diversity of the fauna far exceeds what might reasonably have been predicted if our knowledge of the Cambrian had to rely on normal fossil assemblages consisting of only the remains of skeletons. This, of course, is the usual state of affairs in the fossil record. When palaeontologists collect fossils from an outcrop of Cambrian sediment they can predict with near certainty what the blows of hammer and chisel will reveal. Most abundant will be the remains of trilobites. Very occasionally these fossils will be complete and articulated. More usually, however, the palaeontologist will find only fragments of the exoskeleton. There is also an excellent chance of collecting specimens of brachiopods, maybe some mollusc shells, and perhaps some hyoliths. More rarely other groups may also be found, such as echinoderms and sponges. If a population of these animals happens to be rapidly smothered, then the fossils may occur as articulated remains. How does this compare with the Burgess Shale?

Together Charles Walcott and the Geological Survey of Canada amassed a huge collection of Burgess Shale fossils. It would take months to count every single specimen, but in a shorter time it is possible to obtain some fairly reliable estimates. What such a census[2] reveals is rather significant. First, the existing collections represent approximately 70,000 specimens. Of these, about 95 per cent are either soft-bodied or have thin skeletons, too delicate to survive the normal processes of fossilization. This implies that typical Cambrian assemblages, composed of only skeletal remains, are in comparison with the Burgess Shale very depauperate and represent only about 5 per cent of the individuals alive at any one time (the so-called standing crop). One must admit that the number of species that would be capable of fossilization in ordinary circumstances is somewhat higher, perhaps about 20 per cent. But these figures are very sobering. Applied to other Cambrian assemblages they suggest that a vast amount of information is lost. Second, although greatly impoverished, the shelly remnant from the Burgess Shale would be quite typical of thousands of other Cambrian assemblages. Just like them the residue of shelly taxa from the Burgess Shale consists almost entirely of trilobites, brachiopods, molluscs, and hyoliths. There is an important additional fact. Those species with robust skeletons are not oddities, known only from the Burgess Shale. On the contrary, they are abundant elsewhere and some are widespread, being found not only elsewhere in North America but in other regions such as Siberia. This suggests that if these animals are part of the mainstream of Cambrian life, then it is rather unlikely that the Burgess Shale fauna as a whole represents some strange assemblage stuck in an evolutionary backwater.[3]

The Cambrian 'explosion'

As our knowledge of the Burgess Shale has continued to expand, so it has reopened a whole series of questions that are relevant to the Cambrian 'explosion'. First, we need to understand exactly what is meant by the term 'Cambrian explosion'. The fact that more ancient rocks did not contain obvious evidence for life was articulated at least as early as the 1830s, by the extraordinary and gifted William Buckland, Oxford's first principal incumbent of palaeontology. In his contribution to the *Bridgewater Treatises*[4] Buckland provided some of the first hints of this evolutionary event. By 1859 the problem had been more clearly articulated by Charles Darwin, in his *Origin of species*. Darwin was fully aware that his theory might be difficult to reconcile with the seemingly abrupt appearance of the Cambrian animals.[5] It is a testament both to the prescience of Darwin and the magnitude of the problem that to a considerable extent his articulation of the problem remains compelling and relevant reading today. Since then of course much has been learnt, but only recently has the rather obvious question been asked, at least in detail, as to whether this episode is an 'explosion' in animals or in fossils.[6] In other words is it a genuine evolutionary event or an artefact that appears to be so dramatic because only with the appearance of skeletons could a significant fossil record accumulate.

On balance the evidence suggests that the Cambrian explosion is indeed genuine, but this claim is subject to some important qualifications. It is indeed true that the onset of skeleton formation dramatically improves the quality and extent of the fossil record. As the Burgess Shale fauna graphically shows, the great majority of individuals and even species lack bodies that would be preservable in nearly all environments of deposition. Thus, if an evolutionary radiation were to occur in an entirely soft-bodied group and we had no exceptional fossil preservation, then it would remain undetected and unobserved. There is, however, direct evidence against any such supposition for the Cambrian explosion. Only recently has it been appreciated that the diversification seen in skeletal species is closely paralleled by that of the trace fossils.[7] In most cases the originators of the trace fossils remain unknown, although there are examples where a particular group of animals can be implicated with some confidence. Most notable in this regard are those traces with prominent scratch marks (Fig. 65), which are widely interpreted as representing the digging activity of arthropods. In terms of evolutionary significance, however, the difficulties in establishing correspondence between body and trace fossils is of secondary importance. This is because the diversity of trace fossils emerging in the Cambrian is a clear indication of a dramatic increase in behavioural repertoires and, by implication, of neurological sophistication. Prior to the Cambrian period trace fossils are known, but the earliest abundant examples, which are of Ediacaran age, are substantially simpler.

The Cambrian explosion does therefore appear to be a genuine evolutionary event, and thus one that demands an explanation. Whatever solution is found to this problem need not, of course, embrace the actual origin of

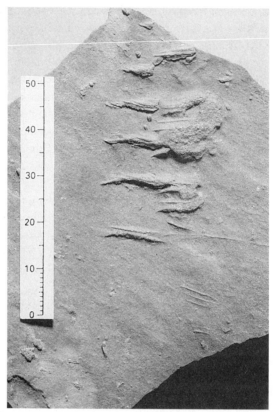

Fig. 65. Trace fossils in the form of a series of scratch marks (*Monomorphichnus*) from the Lower Cambrian Mickwitzia Sandstone of central Sweden. This sequence has a rich record of trace fossils, and is one of the best areas in the world to study their Cambrian diversification. [Photograph courtesy of S. Jensen (University of Cambridge).]

animals. Here the focus of present activity is the attempt to place the Ediacaran faunas into a context that will allow connections to be made with the succeeding Cambrian faunas.[8] This is not proving easy, in part because of the very different styles of fossil preservation. It is also sensible to propose that the earliest fossil animals yet discovered are not the very first animals ever to have existed. In other words there must be some sort of development prior to the Ediacaran history. How deep in geological time this history might have extended is now highly controversial. An independent approach that is attracting wide attention is to look, not at the fossil record, but paradoxically at the molecular similarities of living animals. In essence if two species are very closely related then the sequences that go to build a macromolecule, for example the amino acids in a protein such as haemoglobin, should be very similar, if not identical. Correspondingly two species that shared a common ancestor that lived hundreds or perhaps thousands of millions of years ago will have a markedly different pair of sequences. This, of course, is because once isolated from each other the particular molecule in either of the lineages leading to the two living species will experience repeated substitution along the chain of building blocks that go to make the protein. For example, at one site in the chain there may be the amino acid building block known as alanine. In due course, perhaps a few million years after the lineages first diverged, this amino acid is replaced by another one: say, leucine. The next time this particular site experiences a substitution it may simply revert to alanine, but in general as time elapses the exact sequence of amino acids along the chain will become increasingly different.

It has also been suggested that the substitution of the building blocks of the protein (or DNA) operates at a more or less constant rate during geological time. If this is correct, and it is certainly controversial, then the actual time of divergence between any two species can be estimated by calibrating the observed sequence differences against the known times of divergence in the fossil record. In this way it is possible to establish a sort of 'molecular clock'. In certain cases this 'clock' seems to run at a fairly constant speed, which means that it is then possible to compare molecular sequences in groups of animals with a poor or even non-existent fossil record and so estimate their times of divergence. When such a technique is applied to the animals there is strong evidence, from the molecular clocks, for a rather substantial pre-Ediacaran history.[9] Just how substantial is much more contentious. In one piece of research[10] a variety of protein sequences from several different proteins (and the molecule known as ribosomal RNA) were considered. Each protein therefore provided its own molecular clock. The range of estimates from these clocks for the origins of a number of major groups within the animals was enormous, but nevertheless the figures obtained did consistently point to pre-Ediacaran originations. In an attempt to obtain the best estimate of the actual time of origination the researchers, Greg Wray and his colleagues, averaged the values they obtained and so arrived at a figure for the origin of the animals in excess of 1000 million years. This is substantially older than most earlier estimates. Is it believable? A more critical assessment[11] suggests that averaging is not a very good idea because some clocks seem to run consistently fast and so may be less reliable than the slower clocks. One reason for accepting this assessment is that when the divergence times of vertebrates, which are rather well known from the fossil record, are compared with the values obtained from those molecular clocks that run slow, the correspondence is rather good. Applying such reasoning not just to the vertebrates but to the animals as a whole indicates that the first representatives made their debut about 750 million years ago. This in turn would imply that for the first 150 million years they had no known fossil record. That they had no skeletons may not be surprising, and the general rarity of soft-part preservation may be used to suggest that any discoveries of pre-Ediacaran animals would be highly fortuitous. Such an argument appears to fail, however, because not only are there no obvious body fossils but more import-antly there is also a corresponding lack of trace fossils. This need not mean that there were no animals, but simply an absence of any animals large enough to disrupt or otherwise rearrange the sea-floor sediments. Thus, if they really were present, we can be fairly sure that any pre-Ediacaran animals would have been tiny, only a few millimetres long, and so inhabited a micro-bial world in the benthic realm on the seabed or as floating members of a planktonic community, perhaps similar to some living larvae.[12] What later triggered their initial emergence as the Ediacaran faunas, and subsequently the even more spectacular Cambrian explosion, remains a significant topic for debate.

What we do not know is the extent to which in the pre-Ediacaran faunas—if indeed they ever existed—there were body forms, morphologies, and behaviours that were already present, but could be fully expressed only when some constraining forces (e.g. lack of atmospheric oxygen)[13] were lifted. Alternatively, the first animals may have been for all intents and purposes indistinguishable from the co-occurring eukaryotic microbes, such as a group known as the ciliates. It is certainly difficult to imagine such tiny animals only a few millimetres long being equipped with the full panoply of anatomies that characterize the body plans that are first seen clearly in the Cambrian. The pre-Ediacaran animals would have inhabited a very different world from that of their descendants, and it may be that in these early assemblages there was little in their biology or ecology that has a direct bearing in explaining the Cambrian explosion itself.

Nevertheless, whether we choose to trace our point of initiation as stem-ming from the very first animals, or indeed even further back in the history of life, our present understanding of the Cambrian explosion and the compelling evidence for a profound change in anatomical, ecological, and neurological complexity still forces us to ask some more general questions about the processes of evolution. Ultimately this line of enquiry must pose questions about ourselves and our position in the history of life. After all if, as is patently the case, we are animals and a product of evolution then we need to comprehend our past. Only then shall we be able to understand our present position and perhaps our future. But we are also much more than animals. We need to decide whether evolutionary processes have any bearing on our responsibilities, both to our fellow men and the world we all share.

To embark on an understanding of the Cambrian 'explosion', and let me stress immediately that we are still at a rather preliminary stage of this inves-tigation, we need to combine two main lines of enquiry. First, we should obtain as complete an understanding as is possible of the actual history of events. This in itself is extraordinarily difficult. First, everyone agrees that the fossil record in the Cambrian is seriously incomplete. To start with, not one in a million animals will fossilize. Even with the riches of exceptional preservation from the Burgess Shale, Sirius Passet, and Chengjiang there must be thousands of Cambrian species of which we shall for ever remain in ignorance. To make matters even worse, if the rates of evolutionary change in the Cambrian were very high, then our chances of tracing evolution either within groups or between groups must be reduced. The reason for this is fairly obvious. The likelihood of at least one individual of a species being found is probably more or less constant. If, however, a new group evolves through a series, say ten, of different species in only half a million years, then the chances of finding more than one or two species in this chain of descent are rather unlikely. In fact, the new group will probably appear in the fossil record as if from 'nowhere'. Indeed, palaeontologists acknowledge that such cryptic originations are very common. It is important to understand, however, that this does not necessarily mean that there is some mysterious mechanism of evolution in operation. It is far more plausible to argue that we lack the necessary information.

Nevertheless, despite all these acknowledged problems, there does seem to be a consistency of pattern emerging. Old ideas are modified or even abandoned. New fossils are found, and further discoveries can be expected. If they fit into the present framework, well and good. If not, then we shall have to rethink our hypotheses, perhaps radically. How do we know if we are on the right lines? Some Cambrian fossils do indeed remain very enigmatic. In other instances, however, there does seem to be an internal consistency in our hypotheses. As I shall explain below, this seems to be the case, for example, with the halkieriids and wiwaxiids.

The fossil record by itself, however, will be unlikely to explain all aspects of the Cambrian 'explosion'. That is because this topic is not only a problem for palaeontology: it is a problem for geneticists and evolutionary biologists. In particular, we need to discover by what mechanisms an animal is built. Why does the egg of a fly develop into an insect? What are the basic genetic instructions that determine whether the tissues in the embryo develop into an eye rather than a leg? The explanations lie, of course, in the domain of molecular and developmental biology. It is an area of science that is developing very rapidly indeed. It is also a topic of direct relevance to palaeontologists. If the Cambrian 'explosion' did indeed see an extraordinary variety of animal designs evolving, then perhaps the molecular mechanisms responsible really were different and more potent in their effects than those operating today?

The roots of the Cambrian 'explosion' can almost certainly be traced back to the Ediacaran assemblages. As we saw earlier, there is good evidence that a few animals, notably *Thaumaptilon* (Fig. 33) from the Burgess Shale, represent Ediacaran survivors. Nevertheless, it remains true that the overall differences between the faunas of Ediacaran and Cambrian age are much more striking than any similarities. These differences cannot be simply be explained by the dilution of an Ediacaran component by a crowd of

Cambrian newcomers. Rather, the change that occurred between the two faunas looks much more like a case of replacement. What is not at all clear, however, is whether or not the Ediacaran faunas experienced a dramatic decline before the onset of the Cambrian 'explosion'. In other words, did the Ediacaran faunas plunge into extinction, for whatever reason, so that the Cambrian faunas took the opportunity to occupy a world stripped of its former masters? Alternatively, did the replacement of the Ediacaran faunas take place as a result of the rise of Cambrian animals, the Ediacaran species being extirpated because of bitter competition with the newly evolving animals?

It may be very difficult to decide between these alternatives. In some parts of the world the Ediacaran animals seem to disappear some time before the rise of the Cambrian faunas. This would suggest that the Cambrian 'explosion' was as much an opportunistic event, with vacant ecologies waiting to be reoccupied. Elsewhere, as reported by a team from the Massachusetts Institute of Technology working in Namibia,[14] Cambrian faunas seem to follow almost directly from those of the Ediacaran. This suggests that replacement could have been by competitive interaction. Indeed, the differences between the Ediacaran and Cambrian faunas may transpire to be more apparent than real, and perhaps there are as yet unappreciated aspects of continuity.

Even if the first ripples of the Cambrian 'explosion' are detectable in Ediacaran times, it is still hard to decide which species were crucial to this subsequent evolutionary event as against those that were effectively peripheral. In other words, are there Ediacaran fossils that should be regarded as having a key importance, perhaps central to our understanding of the evolution of a major group? At present, it is very difficult even to begin to answer this question. Although there are a number of Ediacaran fossils that conceivably have such a pivotal status, in all cases their interpretation is very controversial. At the moment most palaeontologists prefer to emphasize the differences between life in Ediacaran as against Cambrian times. For the remainder of this chapter, therefore, the discussion will focus on the evidence from the Cambrian. It is possible that the Ediacaran fossils will come to be regarded as almost irrelevant to how we understand the early evolution of animals. Frankly, I doubt it. In my opinion understanding Ediacaran fossils remains one of the most interesting challenges in palaeontology.

In any event, there does not seem to be any doubt about the magnitude and speed of the Cambrian 'explosion' itself. Because science is all about finding plausible explanations, it is naturally tempting to try to identify *the* trigger that initiated this evolutionary 'explosion': I believe, however, that this approach needs some qualification, especially when we come to consider the nature of organic evolution. History imposes inevitable constraints. At any given stage in the history of life some things, once very likely, may become extremely improbable. Alternatively, a change in the evolutionary situation may unexpectedly facilitate new developments, which in turn open up yet further possibilities. At one level, therefore, the path of evolution must appear to be very unpredictable. As we shall see later, however, at other levels there may in fact be some features in the history of life that are almost inevitable. Another problem in discussing the evolution of life is that it is sometimes surprisingly difficult to talk about triggering mechanisms. This is simply because one event could have occurred only because of some pre-existing state of affairs. But this latter state, in turn, was possible only because of yet earlier conditions, and so on.

Nevertheless, when it comes to understanding the Cambrian 'explosion', it still seems to be desirable to talk about a fundamental trigger that may have initiated the entire process. This is because when animals appeared the world changed, in some ways, for ever. (The same may, incidentally, also be true for the appearance of humans.) In terms of events in the Cambrian (if not the preceding Ediacaran) I believe that the search for the basic trigger may best be sought in the area of molecular evolution. Whatever happened, and some possibilities are discussed below, the consequence was the emergence of new type of organism, latent with new evolutionary possibilities. But how was this potential subsequently realized? In part it almost certainly would have involved further genetic reorganization. Of particular importance, perhaps, is the process known as gene duplication. As the name suggests, an existing gene is doubled up. The old gene continues to act in the usual way, but the new one is potentially free to be employed for new and perhaps unexpected functions. It is also my opinion, however, that whatever the trigger of the Cambrian 'explosion' was, the real motor of this evolutionary event was not so much genetic innovation as an unfolding network of complex and rapidly changing ecological conditions and situations.

The molecular background

First, then, let us consider what might have been the possible molecular background to the Cambrian 'explosion'. Animals, like any organism, are built by a series of instructions that are ultimately transmitted from the genes. Not surprisingly, the basic steps that lead to the construction of an animal are taken at an early stage of development, during the processes of embryology. In the human embryo, for example, one can trace the steps from fertilized cell to a foetus so that in 22 days he or she has a beating heart and by day 32 lenses for the eyes. At the moment our knowledge of what genes are involved with the formation of the muscles of the heart or the cornea of the eye is accelerating at an almost unbelievable rate. It is also clear that if even a very small part of the gene is missing, it may in certain cases lead to severe malformations.

We live in a time of spectacular advances in molecular biology. There is understandably much public interest (and legitimate concern) in areas such as the genetic engineering of crops, counselling for parents concerning the risks in their children of genetically transmitted disorders, and the so-called genome project which aims to document the entire DNA code of humans. There has also, however, been remarkable progress in our understanding of the genetic mechanisms that specify the architecture of animal design.[15] The relevance of this to the Cambrian 'explosion' and faunas such as the Burgess Shale should be obvious. If we can explain how an animal develops from the fertilized egg through a series of embryonic stages in which features such as segmentation and limbs are formed, then there is a fascinating possibility of applying this knowledge to the Cambrian 'explosion'. Do different animals have very different sets of genetic instructions? If so, how might they have evolved and were there special mechanisms operating in the Cambrian evolutionary burst that no longer apply today? Is it necessary to hypothesize a set of genetic instructions that were exceptionally labile, that is, unusually flexible, in order to explain the apparent plethora of animal body plans that irrupted in the Cambrian seas? We are still some way from providing definitive answers, but at the moment, and somewhat surprisingly, the answer to all these questions seems to be 'No'.

The details of the genetic instructions that are needed for the early development of an animal, and hence the establishment of its basic form, are being studied intensively by groups all over the world. Many different genes are being studied, but there is particular interest in a variety of genes that seem to exert crucial roles in the fundamental steps that lead from the fertilized egg to the basic ground plan of the body that includes such features as the distinction between anterior/posterior and dorsal/ventral, as well as segmentation, appendages, and other organs such as the nervous system. Of these genes it is now clear that a class known as the *Hox* genes are especially important. They are known to be involved in very important roles in early development, defining especially the different parts of the body and its overall arrangement as a body plan. They have been studied in considerable detail in arthropods, especially in the fruit-fly *Drosophila*, and in vertebrates, notably the mouse (Fig. 66). The arrangement of *Hox* genes on the chromosome is in a linear array, and it is known that in *Drosophila* different regions of the animal are coded for by specific genes within this elongate genetic complex. What is much more surprising is that there is a direct match between front and back of both the gene complex and the actual body of the animal. In other words, the regions of the fly head are coded for by the genes at the front end of the complex, and so on back through to the tail end of the abdomen. What is even more remarkable is that when the *Hox* complex of the mouse is compared it proves to have the same basic arrangement as is found in *Drosophila*. Crudely, the same genes are involved in the construction of both

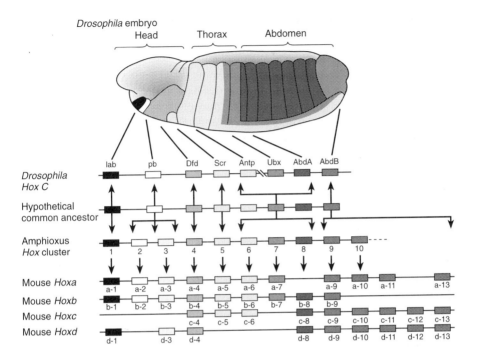

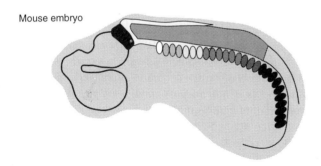

Fig. 66. The distribution of the *Hox* genes in the mouse and fly, and the regions that are coded for by each part of the complex. Despite the obvious differences of the adult, the underlying genetic structure is very similar. [Reprinted with permission from *Nature* (S.B. Carroll. Homeotic genes and the evolution of arthropods and chordates, Vol. 376, pp. 479–85. Copyright (1995) Macmillan Magazines Limited.]

flies and mice. Flies and mice therefore share a basic similarity, and in a certain sense their differences can only be superficial. These animals, of course, differ in all sorts of ways. We would be quite surprised to see a mouse fly past; and even more astonished to meet a fly with whiskers that could suckle its young. There must obviously be other genes coding for the structures that determine why it is the fly and not mice that have wings, and so forth. Nevertheless, the shared presence of these *Hox* genes in flies and mice suggests that their common ancestor, which lived either in the Cambrian or Ediacaran seas, also possessed this genetic complex. You may well be wondering what all this has to do with the Burgess Shale and similar faunas. Here is why I believe that it matters very much. To us, flies and mice look very different; but at a deeper level they have much in common. Both are bilaterally symmetrical, that is having a body divided lengthways into two mirror images, and both share a common ancestor, which lived either in the Cambrian or Ediacaran seas. What did this ancestor look like? Certainly nothing like its flying or furry descendants; it was probably rather similar to the living flatworms (which belong to a phylum known as the platyhelminthes). This worm would also have been bilaterally symmetrical, and we can infer that it would have had certain other features, such as some sort of head containing a brain and also rather primitive eyes. It is now known that the same gene is involved with the coding of the eye in both insects and vertebrates.[16] One can be pretty certain that the Precambrian flatworm also had this gene for its eyes. Since then, of course, much has happened. Each eye of an insect is formed of hundreds of tiny lenses (which make a compound eye) that can probably be processed by the brain to form a sharp image. In contrast, the eye of a mouse is almost identical to ours, with lens, iris, cornea, and so forth.

But how much further back can we trace these *Hox* genes? In fact they are revealing even stranger things. The differences between a fly and a mouse, to our human eyes, are self-evident. At least some of the genes in the *Hox* complex, however, have an even wider distribution among groups of animals. Some, notably the so-called Antennapedia-complex, occur in the cnidarians,[17] which are agreed to be very primitive. This is an important observation for two reasons. First, it hints at the possibility that all animals share a number of basic genetic instructions that could be fundamental to the construction of body plans. In flies and mice the Antennapedia-complex is employed for various purposes, many involved with structures near the front end of the animal. Cnidarians may be primitive in terms of the evolution of animals, but they still have quite a complex organization. They do not, however, have a head, and as yet we do not really know what the Antennapedia-complex does in this group. Conceivably it is involved with helping to determine the principal axis of the animal, differentiating between the end with the mouth, which is surrounded by the tentacles, and the opposite extremity, which is attached to the substratum. Second, the presence of this complex strongly suggests that at least some of the *Hox* genes had appeared very early indeed in the evolution of animals, and certainly no later than Ediacaran times, about 600 Ma ago, when we find the first evidence for fossil cnidarians.

It is beginning to look, therefore, as if many animals may share a fundamentally similar genetic architecture. The recognition of the widespread, and perhaps ubiquitous, presence of the *Hox* genes opens the possibility of think-ing about new ways of defining animals in terms of a common set of develop-mental instructions.[18] It is still important to stress that simply because our bodies (as near relatives of mice) are constructed to the same basic pattern as a fly, in no way does this reduce us to the level of an insect. This point can be made even more forcibly if we compare the genetic make-up and biochem-istry of humans and chimpanzees. In this respect, as is well known, we are very similar indeed. Nearly all the structural genes are indistinguishable, and if one compares the sequence of amino acids that go to form the protein haemoglobin (used for oxygen transport in the blood) it becomes apparent that humans and chimps are identical and do not differ in a single site. This congruence simply reflects our relatively recent divergence from a common ancestor, probably in Africa, and less than ten million years ago. Nevertheless, as I never tire of pointing out to my students in Cambridge, chimpanzees do not play the piano, drink dry martinis, or erect temples to glorify the Creator.

The importance of these observations, therefore, does not directly concern the problem of why certain animals have such anatomical complexity or why one group differs from another. It may turn out that differences that have a self-evident expression in terms of anatomy actually represent rather trivial changes in the underlying genetics, at least so far as the initial stages of diver-gence in the ancestral forms are concerned. What matters in terms of sub-sequent evolution is what the potential consequences of such a change might be. The ancestral animal has, of course, no way of either 'knowing' or being able to guide its evolutionary destiny. The point I wish to stress is that once a certain degree of complexity, such as the evolution of the first animal, is attained, further changes with profound consequences may be facilitated by less dramatic genetic changes.

Animal architecture

Animals, therefore, display an exuberance of design, but appear to have a fundamental similarity at a deeper genetic level. What then actually makes an animal? When one is trying to define something, it may help to say what that thing is not, at least to heighten the contrast. Unfortunately, although we are in the process of uncovering the basic genetic architecture of animals, we are less well placed to discover what might be the crucial genetic differences between animals and their single-celled ancestors belonging to the kingdom of Protista. To tackle this problem we need first to decide which group is most closely related to animals and so shared a common ancestor, pre-sumably in pre-Ediacaran times. Perhaps the most popular suggestion refers to a group of protistans known as the choanoflagellates.[19] These typically consist of small aggregations of cells, which feed in a manner that is very similar to the sponges. Many zoologists believe that aggregation and further developments in complexity, notably some sort of skeletal support and the formation of a series of feeding chambers, in the choanoflagellates led to the first sponges, which are generally agreed to be the most primitive of metazoans.

There are also suggestions that the fungi, familiar to us as the toadstools and mushrooms, are quite closely related to the animals.[20] Again it is very important to stress that if we want to envisage the common ancestor, then it will be rather futile to try to compare a mushroom and a mouse. If fungi and animals do share a common ancestor—and this idea is controversial—then we probably need to consider an organism composed of only a few cells. In any event it is going to be very interesting to see what genetic similarities exist between either the choanoflagellates or fungi and the animals. Perhaps these former groups also possess some of the *Hox* genes?

Animals, however, are obviously different. What then might define an animal? There must be at least one unique feature, more probably several, in the genomic architecture. There are several interesting clues. First, some of the genes involved in the basic organization of animals may have had an early and primitive role in defining body orientation, notably in the specification of the anterior–posterior axis. Most groups of animals have a well-defined symmetry, which is obvious in the bilateral bodies of insects and vertebrates. Another primitive role that seems quite likely is the specification for neural tissue. Neural tissue and the conduction of nervous impulses is another basic feature of nearly all animals.[21] Indeed, the appearance of the nerve cell must be regarded as one of the great steps in the history of life. This is because one path of evolution is then set towards the development of brains, presumably intelligence, and perhaps consciousness. Because at least the first two steps—brains and intelligence—have been acquired at least twice in the history of animals, then an investigation of these similarities between molluscs and vertebrates (and conceivably other phyla) will be rewarding in terms of our evolutionary understanding.[22] Whether the last item, consciousness, will be so amenable is decidedly less clear, at least to this writer.

Is it then possible to identify a genetic trigger connected either to body axes or to neural tissue that was responsible for the appearance of animals and so by implication the Cambrian 'explosion'? It would be tempting, for example, to treat the development of neural tissue as the crucial step; once this evolutionary step was achieved everything else would follow. This may, however, be too sweeping a claim. From our present level of understanding it seems more sensible to identify a series of steps.[23] The first crucial step may have been associated with the synthesis of the special molecules that allow cells to stick to one another. This process, technically referred to as cell adhesion, is a vital prerequisite for any sort of multicellular organism, including the animals, which usually are composed of at least several hundred cells and often millions. It is generally agreed that the most primitive of animals are represented by the sponges. They have several distinct types of cell, but they are not really organized into distinct tissues. Living sponges, however, have no nervous system and at present it seems unlikely that they once had neurons and lost them during their evolution. If we accept sponges as the nearest approximation that we are likely to find to the first animals, then the first important step might have been a type of adhesion that allowed several types of cell not only to stick together, but also to form an organized body.

It is generally agreed that the next stage in the evolution of animals was the appearance of something like a living cnidarian. Here there were probably several crucial steps, the most notable of which would have been the formation of tissues that included a relatively primitive nervous system and the clear definition of body axes. The succeeding steps led to an animal that was probably fairly similar to the living flatworms. The body organization is now structured around a basic bilateral symmetry. The tissues become increasingly complex and include a well-defined nervous system with aggregations of neurons that provide both nerve cords and a brain of sorts.

All these stages had almost certainly been achieved during the Ediacaran interval. Their manifestation, however, is not really apparent until the Cambrian. At which stage the Cambrian 'explosion' became in some sense inevitable is difficult to judge. It does not seem impossible that the evolution of animals could have stopped at the level of organization represented by the sponges, and perhaps even the cnidarians. By the time flatworms had appeared, however, it seems that there could be no turning back. By then the basic genetic architecture was fully in place. Many more changes and innovations were necessary, but I would suggest that it was at this stage that the realm of ecology became the main motor of diversification.

A dangerous world

How the ecology of the Cambrian 'explosion' unfolded is very difficult to model. It must have involved a whole series of complex interactions and feedbacks. But it is still possible that a few crucial factors may be identified. Paramount, perhaps, was the onset and subsequent diversification of those animals that hunt and consume other animals. These are the predators. One might reasonably assume that such evidence would be easy to detect in the fossil record. In reality, it can be surprisingly difficult, as is evident from the fact that for many years it was claimed that Cambrian marine communities were almost entirely free of predators. At this time it was believed that the seas were full of suspension-feeders gently swaying in the sea water and deposit-feeders calmly digging their way through the sediment. This view is now seen to be far too idyllic, but the story of the rise of predators is still quite tentative. It does appear, however, that in contrast to Cambrian communities those of the Ediacaran were largely free of predators. None of the species in the latter communities appears to have possessed a jaw apparatus suitable for seizing and tearing prey, nor is there evidence of damage to fossils consistent with predatory attack. It is certainly possible that the cnidarians had stinging cells similar to those of their living relatives, but whatever prey they captured was probably small. Predators, however, were probably not entirely absent. Sediments of Ediacaran age in central China have yielded tiny calcareous tubes, similar to those found elsewhere in many other parts of the world from rocks of the same age. Unusually, however, the Chinese fossils, which belong to a genus known as *Cloudina*, have tiny boreholes.[24] Unfortunately, we have no good idea what sort of predator was able to bore into the tubes, presumably to suck out the soft tissues. Animals as primitive as flatworms are known to make boreholes in their prey, so here is one possibility.

It has long been appreciated that one of the main functions of external skeletons in the many animals that possess them is to provide protection from attack. It requires no great leap in imagination, therefore, to link the abrupt appearance of skeletons in the Cambrian to the introduction of predators. It is well to remember that animals can protect themselves in all sorts of ways and may employ devices such as toxins, warning coloration, and camouflage. Some scientists have thought that invoking skeletons as protective armour is somewhat naïve. It is agreed that in some cases they have additional functions, such as providing support for soft tissues. It may also be rather unwise to extrapolate what we see among living marine animals back into the Cambrian: the intensity and sophistication of attack and deterrence in modern-day oceans may far outstrip what occurred half a billion years ago.

Despite all these provisos, the idea that the primary role of skeletons is protective now looks likely, even though once acquired these hard parts may well have conferred all sorts of other advantages and opportunities. The evidence to support the defence hypothesis comes from the arrangement of the skeletons themselves and a growing list of examples of attack. In terms of skeletal architecture, perhaps the most notable examples are from the cataphract (chain-mail-like) covering of animals such as the halkieriids (Fig. 67). Their scleritome is superbly engineered, in that the closely packed sclerites are so arranged to provide a coherent but flexible covering.[25] A number of other animals, such as the Burgess Shale *Hallucigenia* (Figs 18, 19) and *Wiwaxia* (Fig. 43), carry strikingly elongate spines whose primary function was surely protective. In some specimens of *Wiwaxia* one or more spines appear to have been snapped off, presumably by unsuccessful assailants.

Skeletons could not, of course, provide complete immunity, and apart from breakage of protective spines there is convincing evidence for attack from

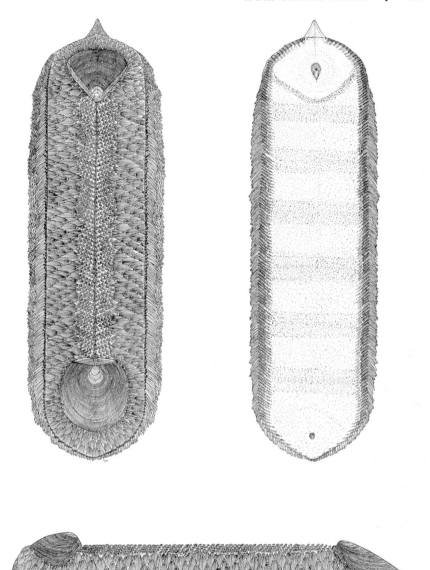

Fig. 67. Reconstruction of the halkieriid *Halkieria evangelista* from the Sirius Passet fauna of North Greenland. Left, dorsal; right, ventral; lower, lateral. [Reproduced with permission from fig. 2 in S. Conway Morris and J.S. Peel (1995). Articulated halkieriids from the Lower Cambrian of North Greenland and their role in early protostome evolution. *Philosophical Transactions of the Royal Society of London* B, Vol. 347, pp. 305–8.]

several other sources. The most dramatic evidence comes from the trilobites, where bite marks are quite frequent (Fig. 41). It is likely that in many of the fossils the attack was not lethal and the trilobite survived. In some cases there is clear evidence of healing, and it is possible that the animal had efficient wound-repair mechanisms, similar to those employed by living arthropods. The trilobites that appeared to have survived attack have mostly bite-marks on the sides or back of the skeleton. Such wounds would have been serious, but would have tended to avoid vital organs. In contrast, evidence for bite-marks on the head region is much rarer. Here death may have been almost inevitable.

There is one rather extraordinary feature of the bite-marks in Cambrian trilobites that was noted by the American palaeontologist Loren Babcock.[26] If a census is made of those on the right-hand side of the animal as against those on the left-hand side, one might predict more or less equal numbers. Rather remarkably there is a marked preponderance of dextral attacks. This bias, which is an example of what is known as laterality, is familiar to us because of the preponderance of right-handed humans. Its presence in the Cambrian is perhaps more surprising. It suggests that either the attacker had a dextral preference or there was a bias in the direction by which the trilobite tried to escape.

Because trilobites tend to be the most abundant Cambrian fossils and because many of the other groups with skeletons are often rather small, if not microscopic, it is not too surprising that most of the available evidence for predation comes from trilobites. Boreholes are, however, known from some brachiopods (Fig. 68).[27] Interestingly some of these are incomplete, representing abortive attempts at drilling that were perhaps abandoned, either because the shell proved too thick and resistant or because the attacker was disturbed. The brachiopods themselves are rather small, normally less than a centimetre across. The boreholes themselves are tiny, typically about 0.2 mm in diameter. As with the rare examples in the tubes of *Cloudina* from the Ediacaran of central China, the nature of the assailant is a matter for speculation.

The initial realization of the importance of predation in the Cambrian stemmed almost entirely from the research into the Burgess Shale. With these studies it soon became clear why hitherto the Cambrian seas had been thought to be very largely free from predators. This was because much of the necessary evidence came from fossils with a minimal fossilization potential. As we saw above, there is compelling evidence for predation in the Burgess Shale. But recall the main players. There were the hyoliths swallowed by the priapulid *Ottoia*, the long nozzle-like feeding extension of *Opabinia* ceaselessly seeking out its prey of small worms, and most notable of all the great *Anomalocaris* with its giant grasping appendages and strange diaphragm-like jaw. The crucial point about all these animals is that their likelihood of surviving in the normal circumstances of fossilization is low. Admittedly *Anomalocaris* had an apparently tough exoskeleton, but because it was not impregnated with mineral salts in the way that a trilobite is, its chances of

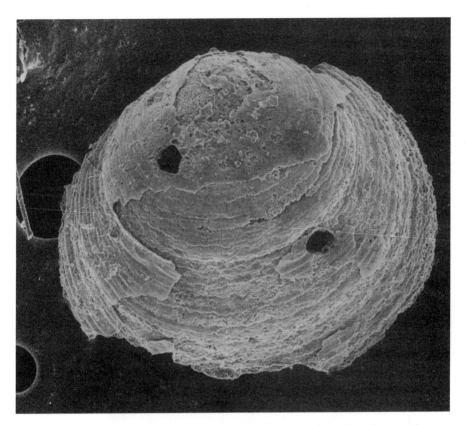

Fig. 68. A Middle Cambrian brachiopod, from southern Sweden, with two prominent holes drilled by an unknown predator.

entering the fossil record are low. It is true that *Anomalocaris* is relatively widespread as a Cambrian fossil, but the only portion that usually survives is the giant frontal appendage. Recall that this was interpreted first as an abdomen and then as a walking leg. If the complete specimens of *Anomalocaris* had not been found it is very unlikely this animal would have ever been identified as a predator.

We can, of course, only infer the presence of animals like *Anomalocaris* and *Ottoia* in the great majority of Cambrian communities. But this inference is not an unreasonable assumption. In addition, with careful searching more examples of bitten trilobites and brachiopods with boreholes will no doubt be recognized. It is important to remember, however, the examples of the hyoliths within the intestine of *Ottoia*.[28] So far as we can be tell, even though the soft tissues within the hyolith shell would have been digested by the gastric juices of the priapulid worm, the shells themselves would have emerged intact with no indication that they had passed through the gut of a predator.

Where else may we look for evidence for predation? One intriguing line of evidence concerns the depth of burrowing of trace fossils. In Ediacaran faunas traces are quite common, but as noted above (p. 30) they tend to follow a single horizon within the sediment, and are not drilled vertically. In contrast, in the Cambrian not only do trace fossils become more diverse, but for the first time we find abundant signs of vertical burrows. In Cambrian sands that accumulated in shallow water, close to the coast, these vertical burrows are so abundant that the sediment has been termed 'pipe-rock' (Fig. 69).[29] It is also apparent that in the Cambrian there is a general increase in the intensity of churning of the sediment (a process known as bioturbation). Not only that, but the overall depth to which this bioturbation extends also increases through time.[30] It seems possible that the onset of vertical

Fig. 69. Vertical burrows of *Skolithos* in a Cambrian sandstone from southern Sweden. It is not known what sort of animal lived in these burrows. Such a style of burrowing is very unusual in Ediacaran sediments, but is widespread in the Cambrian. Their abundance leads to the term 'pipe-rock'.

Fig. 70. Two trace fossils. The large one represents scratch marks made by a large arthropod. They intersect the burrow made by a worm. It looks as if the arthropod hunted for and then ate the worm. [Photograph courtesy of S. Jensen (University of Cambridge).]

burrowing and the increasing depths of bioturbation are in part a direct response to rising levels of predation, with soft-bodied animals seeking refuge within the seabed. Nowhere, however, was safe. Just as hard skeletons cannot confer complete immunity, so the Cambrian hunters would have followed their prey into the sediment. Dramatic evidence for this comes from Lower Cambrian sandstones of central Sweden.[31] Some of the trace fossils represent large excavations, scratched into the sediment by large arthropods. Similar traces are widespread, but these Swedish examples are rather unusual because very often the excavation intersects the burrow of a worm (Fig. 70). There is little doubt that the arthropod was a predator, and the worm the prey.

The unfolding ecological theatre

It is one thing to show that predation was an important factor in Cambrian ecosystems, but it is quite another to show that in the absence of predators either the Cambrian 'explosion' would not have happened, or it would have been a much slower process. The reason for this is that even in modern communities the role of predation remains controversial. Nevertheless, it seems likely that in some cases the presence of predators is a vital component in the maintenance of ecological richness and biological diversity in living communities. At first sight this observation appears to be paradoxical, in that one might intuitively regard predators as destructive agents. Simple experiments in the natural environment, however, suggest otherwise.[32] If the dominant predator is removed, not surprisingly there is a population explosion among the various prey animals as the ecological pressure is released. In many circumstances, however, one species quickly rises to dominance, overwhelming and smothering the others. The net result is a crash in diversity, leaving a depauperate and dull community. If the predator is then reintroduced, then its apparently disruptive action will allow diversity to climb back towards its original levels. In some wider sense, so it is proposed, the rise of Cambrian predators may have helped to drive forward the 'explosion' in diversity.

It may also be the case, however, that the onset of other types of feeding among Cambrian animals had equally important effects on the ecology and so on Cambrian diversification. Of particular importance is the style of feeding known as grazing. The concept of grazing is often exemplified by animals such as slugs and snails. Their feeding apparatus consists of a specialized rasping structure (the radula). This is protruded through the mouth to scratch away thin films composed of algae and bacteria. In marine environments, however, grazing encompasses a number of other groups, notably the sea urchins (echinoids) and various molluscs, notably some snails and the less well-known chitons. What about Cambrian grazers? First there was a wide variety of molluscs. Some would have been deposit-feeders, seeking out grains of sediment and organic fragments. Other molluscs, however, appear to have been grazers, as do the halkieriids (Figs 53, 67) and the related *Wiwaxia* (Fig. 43). The influence of grazing on the Cambrian ecologies may well have been analogous to that of predation. Just as the primary role of animal skeletons was to provide a protective shield, so, it has been argued, the extensive development of calcareous deposits in different sorts of algae was an attempt to cushion the effects of grazing.[33]

So far I have been considering the type of ecology that largely impinged on or in the sea floor. What was happening above in the pelagic realm? There is certainly some evidence for the activities of predators, notably in the form of the Burgess Shale animal *Nectocaris* (Fig. 49) and the chaetognaths. A very important aspect of pelagic ecology, however, is the harvesting of the microplankton by animals that strain or sieve the sea water. Because much of the microplankton lacks skeletons or other preservable hard parts, the original composition and diversity of these communities in the fossil record may be difficult to judge. Some plankton, however, do have preservable remains. In oceans today the protistan plankton includes groups such as the coccolithophorids and dinoflagellates. In the Cambrian, however, the equivalent assemblages are in general difficult to assign to specific groups. Instead they are referred to as the acritarchs, although this is little more than a 'dustbin' term. Despite these uncertainties the history of the acritarchs is becoming better known. They first appear far back in the Precambrian, more than a billion years ago. Over time they become moderately diverse, but during a series of major ice ages that precede the onset of the Ediacaran faunas (Fig. 3) the acritarchs crash in diversity as they experience major extinctions.[34] Thus, during the Ediacaran interval, about 560 Ma ago, the acritarch assemblages are rather depauperate. In the Cambrian, however, they become increasingly diverse, many showing spines or other ornamentation. The acritarchs therefore experience their own Cambrian 'explosion'. Is there any link between this event and the rise in animal diversity?

The best evidence for such a link may come from the wonderfully preserved arthropods recovered from the Mount Cap Formation of north-west Canada. The fortuitous discovery of these fossils by Nick Butterfield[35] was explained in Chapter 4. Although most of the fossils are fragmentary, the quality of preservation is almost unbelievable (Fig. 58) and the specimens could quite easily be mistaken for modern material. But there is no evidence for inadvertent contamination. Various pieces of arthropod were recovered by Nick Butterfield. Some of the most interesting were elongate appendages, which bore numerous filamentous extensions. These appendages may occur in parallel arrays, so that the fossil presents the general appearance of an extremely effective sieve. And this appears to have been their function. Presumably the original arthropods, of which we know rather little, because so far only fragments have been found, were closely similar to the living crustaceans that still scull through the oceans, ceaselessly combing the sea water for suspended food particles. In the Cambrian seas the food of these arthropods would have included the acritarchs. Nick Butterfield has speculated that the observed increase in diversity of acritarchs during the Cambrian and the related development of projecting spines or other ornamentation is a direct consequence of the grazing pressure exerted by these filter-feeding activities of the arthropods. This is a very intriguing idea.[36] Nevertheless, it must be admitted that we know very little about the functional significance of acritarch design, nor what relationship it bears to the feeding activities of arthropods.

The possibility that the Cambrian 'explosion' can be understood, at least in part, by changes in food sources and their exploitation by the newly evolving animals is by no means a new idea. Some years ago the distinguished American palaeontologist Jim Valentine suggested that there were significant changes in the stability of food supply ('trophic resources' is the term used in biology) across the Precambrian–Cambrian boundary.[37] In his opinion these changes might have been responsible for driving forward the diversification of animals, as well as other features, such as the increasing degree of sediment disturbance, including the widespread appearance of the vertical burrows. Why should trophic resources change in this way? Jim Valentine proposed that the underlying reasons be sought in the mechanisms of plate tectonics and continental drift. How reasonable is this? There is certainly evidence that in late Precambrian times many of the present-day continents were welded together into what geologists call a super-continent. Subsequently this began to break up into a number of continents, each separated by seaways or even oceans. Overall, of course, the distribution of these continents was very different from that of today. Where I write, in Cambridge, appears to have been quite close to the South Pole about 700 Ma ago, while other areas such as the islands that now form Japan had not even come into existence. The ancient continent of Laurentia, which is the main repository of Burgess Shale-type faunas, was also removed from its present position and in the Cambrian straddled the equator (Fig. 50).

It is certainly not immediately clear how this break-up of the super-continent might change the balance and distribution of food supply in the oceans. But perhaps there are some clues. The interval marked by the Cambrian 'explosion' is also a time when quite unusual amounts of sedimentary phosphate were being deposited in the shallow shelf seas that rimmed continents. These regions are now found in such places as Australia, South China, and Kazakhstan.[38] Precise estimates of the total volume of phosphate that accumulated in the geological interval are not easy to obtain. Nevertheless, the fact that some of the world's most important mines that extract this phosphate, largely to provide agricultural fertilizer, are situated in rocks of Cambrian age gives a crude indication of the massive quantities of phosphorus that must have been deposited. Unfortunately, the mechanisms of the accumulation and preservation of phosphate-rich sediment are decidedly controversial. There is little doubt that special conditions must have been present in the oceans, but it is far from clear whether the concentrations of phosphorus in Cambrian sea water itself were in fact elevated. Some geologists believe that there could be a direct correlation between the Cambrian 'explosion' and this episode of phosphogenesis. Quite how elevated levels of phosphorus might have helped to drive Cambrian evolution is, nevertheless, not very obvious. Nobody denies the importance of phosphorus. It is a vital nutrient for life. Its enhanced abundance in areas of oceanic upwelling today, such as off the west coast of South America, is of great importance, not least for the economics of fisheries. It is much less clear what influence enhanced nutrient supply and productivity might have had on the Cambrian 'explosion'. In other words, creating an abundance of life and a huge biomass by the presence of high concentrations of phosphorus does not in itself seem to guarantee rapid rates of evolution.

Despite all these uncertainties there does seem reason to believe that the unfolding of the Cambrian 'explosion' was largely governed by a series of ecological feedbacks. Some workers, notably Mark McMenamin,[39] have thought that this feedback must have been an unpredictable and essentially chaotic process. I believe, to the contrary, that this is an exaggeration, and that such an approach might obscure some underlying patterns. In outline, the methods of obtaining and processing food are not unlimited. Neither are other activities of animals, such as locomotion or anchorage to the sea floor, that have an immediate bearing on their ecology. Thus, although it may indeed be difficult to establish which groups of animals rose to ecological success in the Cambrian explosion, the constraints of ecology nevertheless suggest that once the range of anatomies is established as part of the Cambrian explosion then it should be possible to investigate causations and thereby confer some sort of predictability. One of our principal conclusions is that the role of predation was very important and had at least one direct feedback; the promotion of protective skeletons. Quite possibly other strategies for evading attack, including deeper burrowing, were also a consequence of predation. The activities of grazers and filter-feeders may have been just as important,[40] but as yet the nature of the feedback mechanisms is not so obvious.

Study of the Burgess Shale itself reveals a complex ecology. It remains imperfectly understood because, unless we really do invent time travel, even this superb fauna cannot reveal a complete set of insights into Cambrian life. In general, the ecology of the Burgess Shale looks remarkably modern, with well-defined groups of suspension-feeders, deposit-feeders, and carnivores, all linked by a complex food web. In this and similar faunas it seems likely that different ecological niches were subdivided to a considerable degree. Direct evidence is, however, quite difficult to obtain. Some years ago ecologists studying living communities were interested in trying to describe the structure of ecological niches and the distribution of resources in a biological community by using mathematical models.[41] One way of visualizing such models is to consider the distribution that arises when the abundance of each species, measured in terms of numbers of individuals (or alternatively biomass), is plotted against their overall rank abundance (1 = the most abundant, 2 = the second most abundant, and so on). In general there appear to be three main distributions, each of which can be linked to a particular model of resource distribution (Fig. 71). Because of the abundance and relative completeness of the Burgess Shale fauna, such analyses can also be applied to this community.[42] For the most part, the ecological categories recognized fall into a distribution known as log-normal. The significance of this distribution is still being disputed by ecologists, although some suggest that the best interpretation is that it represents evidence for a stable ecological system. In the Burgess Shale, however, there was at least one interesting exception. Those animals, such as *Marrella* (Fig. 16), that are identified as mobile deposit-feeders, living on rather than in the sea floor, fall into another type of distribution, known as geometric. In this case the most abundant species takes a given fraction of the total resource, say 40 per cent. The next most abundant species (number 2 in rank order) takes 40 per cent of the remainder, and so on.

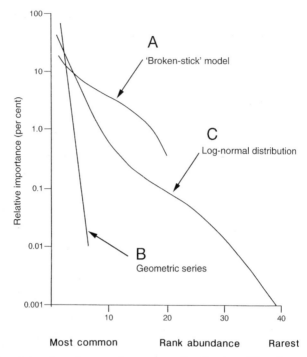

Fig. 71. Models of ecology and resource distribution. The three principal distributions of taxa in terms of relative importance plotted against rank abundance. *A* is the so-called 'broken-stick' model; *B* refers to the geometric distribution; *C* is an example of a log-normal distribution. Each distribution has been linked to a number of ecological models. [Redrawn from Fig. 2.10 of R.H. Whittaker. *Communities and ecosystems.* © Copyright, 1970. The Macmillan Company, New York and Prentice Hall.]

Understanding the complexities of the Cambrian 'explosion' in terms of its ecology remains a provocative and challenging prospect. It is now necessary, however, to consider the other side of the Cambrian coin. Overall, even if the role of ecology in the Cambrian is still poorly understood, the actual types of ecology do not appear to be radically different from those obtaining among marine animals today. In many instances, however, the Cambrian animals themselves appear very different. Indeed, how can we explain the origin of so many forms of animal organization in the Cambrian? Did they appear almost simultaneously? Surely there must have been extraordinary mechanisms of evolution responsible for this, mechanisms that no longer operate today? Perhaps even the Darwinian paradigm of evolution, vigorously defended against all attack for more than a century, is now set to crumble before our eyes? Is not the sheer range of animals alive in the Cambrian far in excess of anything we see in modern seas? These are not absurd suggestions made by cranks and eccentrics on the fringes of science, but questions raised by reputable scientists. Unfortunately, however exciting these revolutionary ideas might appear to be, in fact they appear to be deeply flawed. In one way the reality of the Cambrian explosion is much more sobering and mundane. In the next chapter we shall see exactly why.

Notes on Chapter 6

1. Full details are: S.J. Gould. 1989. *Wonderful life. The Burgess Shale and the nature of history*. Norton, New York.
2. My census of the Burgess Shale and its analysis was published in *Palaeontology* (Vol. 29, pp. 423–67 [1986]). Reconsidering the data that I accumulated, I think it likely that I underestimated the standing crop of trilobites, because apart from *Olenoides*, of which the associated soft parts are quite frequently preserved, for the other ptychopariid trilobites I included in the census only the very rare specimens with some evidence for soft parts. There are, however, several hundred articulated specimens, and these I now believe were alive at the time of burial. This means that the shelly component of the Burgess Shale fauna would have made a slightly larger contribution to the standing crop than I originally calculated. In addition there have been some minor changes in the taxonomy.
3. Similar arguments can be applied to the Chengjiang fauna, which houses an assemblage of 'normal' Cambrian trilobites and brachiopods. A curious absentee appears to be the echinoderms, although they are also rare in the Burgess Shale. So far as I am aware no detailed census of the Chengjiang fauna is available, although it is clear that as with the Burgess Shale the bulk of the fauna is composed of only a few species. The question of whether the Sirius Passet fauna is representative of Cambrian sea-floor communities is, however, more interesting. Here the shelly fauna is markedly depauperate, with only one species of trilobite, minute brachiopods, and apparently neither molluscs (excluding rare hyoliths) nor echinoderms. The soft-bodied component is, however, quite diverse. One possible explanation for the differences between the Burgess Shale and Chengjiang faunas (which are quite similar yet separated in geological time) and the Sirius Passet fauna is that the last-named inhabited an area of seabed with markedly lower oxygen concentrations.
4. See the two volumes by W. Buckland entitled *Geology and mineralogy considered with reference to natural theology*. (Pickering, London, 1836). This was one contribution to the famous set of Bridgewater Treatises, the authors for which were specifically commissioned to present evidence for the power, wisdom, and goodness of God as manifested in the Creation.
5. A substantial part of Chapter 9 of *On the origin of species* addresses this problem.
6. An important paper in this regard is by B. Runnegar in the *Journal of the Geological Society of Australia* (Vol. 29, pp. 395–411 [1982]).
7. Useful reviews in this area by T.P. Crimes may be found in a series of papers, including those in *Journal of the Geological Society, London* (Vol. 149, pp. 637–46 [1992]) and in *The palaeobiology of trace fossils* (ed. S.K. Donovan), pp. 105–33 (Wiley, 1994). The relative abruptness and scale of the diversification depends in part on the richness of the trace-fossil record of Ediacaran age. Soren Jensen informs me that a number of purported Ediacaran trace fossils require renewed scrutiny, and that his research is pointing towards a major shift in trace-making activity from near the beginning of the Cambrian.
8. Because of the popularity of the Vendobionta hypothesis (see Chapter 2) there has been less interest in attempting to accommodate the Ediacaran fossil record into the framework of metazoan evolution. I tried to present the beginnings of an overview in an article in *Nature* (Vol. 361, pp. 219–25 [1993]). My

incorporation of the Ediacaran fronds and Burgess Shale *Thaumaptilon* into the pennatulaceans in *Palaeontology* (Vol. 36, pp. 593–635 [1993]) complements earlier work by M.F. Glaessner (*The dawn of animal life: a biohistorical study* (Cambridge University Press, 1984), R.J.F. Jenkins (in *Origin and early evolution of the Metazoa* (ed. J.H. Lipps and P.W. Signor), pp. 131–76 (Plenum Press, New York, 1992), and J.G. Gehling (*Memoirs of the Geological Society of India* (Vol. 20, pp. 181–224 [1991]), who have consistently argued for metazoan affinities of Ediacaran fossils. More recently the cudgels have also been taken up by B.M. Waggoner in *Systematic Biology* (Vol. 45, pp. 190–222 [1996]).

9. Two key papers in this regard are by B. Runnegar on the molecular clocks of the proteins haemoglobin (*Lethaia*, Vol. 15, pp. 199–205 [1982]) and collagen (*Journal of Molecular Evolution*, Vol. 22, pp. 141–9 [1985]).

10. The paper by G.A. Wray and colleagues was published in *Science* (Vol. 274, pp. 568–73 [1996]).

11. See my commentary in *Current Biology* (Vol. 7, pp. R71–4 [1997]), and also that by M.A. Bell (*Trends in Ecology and Evolution*, Vol. 12, pp. 1–2 [1997]).

12. Both these ideas have received extensive consideration. The idea that the earliest metazoans were comparable to the present-day meiofauna, the extraordinary set of miniaturized metazoans that inhabit the interstices of sandy sediments, was vigorously championed some years ago by various workers such as P. Boaden (*Zoological Journal of the Linnean Society*, Vol. 96, pp. 217–27 [1989]), but more recently has received little support. The other idea that larval morphologies are an important ingredient in understanding the early stages of metazoan evolution has been revitalized by E.H. Davidson and others in an article in *Science* (Vol. 270, pp. 1319–25 [1995]).

13. The possible role of an increase in atmospheric oxygen providing an impetus for the Cambrian explosion has been a perennial favourite. Recently some geochemical data has helped to fuel, so to speak, the argument with important papers in *Nature* by G.A. Logan and others (Vol. 376, pp. 53–6 [1995]) and by D.E. Canfield and A. Teske (Vol. 383, pp. 127–32 [1996]). A short commentary by A.L.R. Thomas in *Trends in Ecology and Evolution* (Vol. 12, pp. 44–5 [1997]) is also relevant.

14. The relevant paper is by J.P. Grotzinger and others in *Science* (Vol. 270, pp. 598–604 [1995]).

15. There is a huge and burgeoning literature that makes it difficult to keep abreast of this fast-moving field. A succinct and readable summary of the present position of developmental biology, evolution, and palaeontology is the book by R.A. Raff entitled *The shape of life: genes, development, and the evolution of animal form* (Chicago University Press, 1996). Other accessible summaries are those by M. Akam (*Philosophical Transactions of the Royal Society of London* B, vol. 349, pp. 313–19 [1995]), M. Averof and others (*Cell and Developmental Biology*, Vol. 7, pp. 539–51 [1996]), C. Kenyon (*Cell*, Vol. 78, pp. 175–80 [1994]), S.B. Carroll (*Nature*, Vol. 376, pp. 479–85 [1995]), and P.W.H. Holland and J. Garcia-Fernandez (*Developmental Biology*, Vol. 173, pp. 382–95 [1996]). In addition, the 1994 Supplement of the journal *Development*, edited by M. Akam and others, contains a whole series of relevant papers.

16. See note 8 in Chapter 1.

17. Key references to cnidarian *Hox* genes may be found in papers by B.L. Aerne and others (*Developmental Biology*, Vol. 169, pp. 547–56 [1995]), K. Kuhn and others (*Molecular Phylogeny and Evolution*, Vol. 6, pp. 30–8 [1996]), and B. Schierwater and others (*Journal of Experimental Zoology*, Vol. 260, pp. 413–16 [1991]).

18. This idea has been articulated as the phylotypic or zootype concept, and is explained by J.M.W. Slack and others in *Nature* (Vol. 361, pp. 490–2 [1993]).

19. An introduction to the choanoflagellates, and indeed the range of animal phyla,

can be found in the book by C. Nielsen entitled *Animal evolution: inter-relationships of the living phyla*. (Oxford University Press, 1995).

20. The idea of a metazoan–fungal relationship is based on evidence from molecular biology. Papers that discuss these data include those by S.L. Baldauf and J.D. Palmer (*Proceedings of the National Academy of Sciences, USA*, Vol. 90, pp. 11558–62 [1993]) and P.O. Wainwright and others (*Science*, Vol. 260, pp. 340–2 [1993]). A recent rather sensational discovery that may support this supposition is the recognition of collagen in fungi, a structural protein that hitherto had thought to be restricted to the metazoans (see M. Celerin and others in *EMBO Journal*, Vol. 15, pp. 4445–53 [1996]). An alternative possibility is that the shared possession of collagen represents evolutionary convergence rather than a phylogenetic relationship. A more sceptical view of a relationship between metazoans and fungi is expressed by A.G. Rodrigo and others in *Systematic Biology* (Vol. 43, pp. 578–84 [1994]).

21. The exception is the sponges, widely regarded as the most primitive of metazoans. Although there is evidence for communication across the sponge there is no indication that it is mediated by nervous tissue (see T.L. Simpson's book *The cell biology of sponges* (Springer Verlag, Heidelberg, 1984)).

22. Although the intelligence of molluscan cephalopods, especially the octopus, have been long appreciated, detailed information on the complexity of the brain is only now emerging (see the paper by B.U. Budelmann in *The nervous system of invertebrates: an evolutionary and comparative approach* (ed. O. Breidbach and W. Kutsch), pp. 115–38 (Birkhäuser Boston, Cambridge, Massachusetts, 1995).

23. The paper by D. Erwin published in *Biological Journal of the Linnean Society* (Vol. 50, pp. 255–74 [1993]) gives a valuable discussion of these topics.

24. This, the earliest evidence for predation in the fossil record, is reported by S. Bengtson and Z. Yue in *Science* (Vol. 257, pp. 367–9 [1992]).

25. The importance of the halkieriids is returned to in the next chapter. Information on the articulated material was published by myself and J.S. Peel in *Philosophical Transactions of the Royal Society of London* B, Vol. 347, pp. 305–58 [1995]).

26. See the paper by L.E. Babcock in *Journal of Paleontology* (Vol. 67, pp. 217–29 [1993]).

27. Evidence for borings in Middle Cambrian brachiopods was published by me and S. Bengtson in *Journal of Paleontology* (Vol. 68, pp. 1–23 [1994]).

28. The details of the hyoliths in the gut of *Ottoia* were published by me in *Special Papers in Palaeontology* (Vol. 20, pp. i–iv, 1–95 [1977]).

29. The distribution of 'pipe-rock' through geological time is reviewed by M.L. Droser in *Palaios* (Vol. 6, pp. 316–25 [1991]). It is a curious fact that sediments with 'pipe-rock' are very largely a phenomenon of the Lower Palaeozoic, and especially the Cambrian, where this fabric is very common. The reasons for its virtual disappearance later in geological time are uncertain. They may include the relative restriction of near-shore sands, in which the vertical burrows termed *Skolithos* occur. More plausibly, the style of bioturbation and the increasing degree of disturbance may militate against the survival of 'pipe-rock' in most circumstances.

30. The increasing levels of bioturbation during geological time have been reviewed by M.L. Droser and D.J. Bottjer in *Annual Review of Earth and Planetary Sciences* (Vol. 21, pp. 205–25 [1993]).

31. The relevant paper is by S. Jensen in *Lethaia* (Vol. 23, pp. 29–42 [1990]). In this paper Soren Jensen speculates that the victim of the predator and maker of the burrow was a priapulid. The priapulids are of course the most conspicuous of the worms in the Burgess Shale.

32. Classic experiments in this regard were conducted by R.T. Paine on the communities of rocky coastlines. An overview is given in *Paleobiology* (Vol. 7, pp. 553–60 [1981]).

33. It has long been recognized that the first major onset of calcification in a number of different groups of algae, including both prokaryotic cyanobacteria and various eukaryotes, occurs in the Cambrian. A useful introduction to this area is by R. Riding and L. Voronova in *Geological Magazine* (Vol. 121, pp. 205–10 [1984]). It should be pointed out that the evidence for calcification of microbes in Precambrian sediments is growing (see the paper by A.H. Knoll and others in *Palaios* (Vol. 8, pp. 512–25 [1993]). It is also necessary to stress that the problems of calcification in algae are by no means resolved, and that calcification in part may be controlled by the degree of carbonate saturation of the oceans.

34. An update of protistan diversity during the Proterozoic and Cambrian is given by A.H. Knoll in *Proceedings of the National Academy of Sciences, USA* (Vol. 91, pp. 6743–50 [1994]).

35. The paper describing these fossils was published by N.J. Butterfield in *Nature* (Vol. 369, pp. 477–9 [1994]).

36. There is also perhaps a link to an ingenious hypothesis, based on geochemical data, by G.A. Logan and others (*Nature*, Vol. 376, pp. 53–6 [1995]) to the effect that the development of planktonic grazers and the rapid transfer of organic matter to the seabed via faecal pellets as against its slow descent as a sort of marine 'snow' led to profound changes in ocean states, notably the degree of oxygenation.

37. A summary of these ideas can be found in the paper by J.W. Valentine in *American Zoologist* (Vol. 15, pp. 391–404 [1975]).

38. The preponderance of phosphatic deposits across the Precambrian–Cambrian boundary has been widely commented upon. An overview can be found in several publications including *Phosphate deposits of the world, Volume 1. Proterozoic and Cambrian phosphorites* (ed. P.J. Cook and J.H. Shergold) (Cambridge University Press, 1986), and by M.D. Brasier in *Geological Society of London Special Publications* (No. 52, pp. 289–303 [1990]) and in *Origin and early evolution of the Metazoa* (ed. J.H. Lipps and P.W. Signor), pp. 483–523 (Plenum Press, New York, 1992).

39. See M.A.S. McMenamin and D.L.S. McMenamin, *The emergence of animals: the Cambrian breakthrough* (Columbia University Press, New York, 1990). For a critical review, which I would endorse, see R.A. Fortey's trenchant comments in *Historical Biology* (Vol. 4, pp. 70–1 [1990]).

40. This area has been developed by N.J. Butterfield in *Paleobiology* (Vol. 23, pp. 247–62 [1997]).

41. This area of mathematical ecology has spawned a large literature. The summary by R.M. May in *Ecology and evolution of communities* (ed. M.L. Cody and J.M. Diamond), pp. 81–120 (Belknap Press, Harvard, 1975) is especially useful.

42. This work was published by me in *Palaeontology* (Vol. 29, pp. 423–67 [1986]).

CHAPTER 7

Animal architecture and the origin of body plans

Introduction

Many of the animals from the Burgess Shale and similar deposits look, to our eyes, to be very peculiar, if not downright bizarre. It has been thought that many of these animals were so odd, so different from anything we know, that there was no possibility of accommodating them in any known phylum. This is an important claim because the concept of the phylum is generally taken to be basic to our understanding of animals inasmuch as each of the phyla corresponds to one of the 35 or so basic body plans identified today (see also Glossary). The notion of a body plan, and its correspondence to the phylum, is a very useful and popular concept in trying to bring some order to the complexities of animal classification. It has, however, some hidden pitfalls. In part this is because of a tension, largely unacknowledged, between a decidedly static concept of a body plan, sometimes referred to as an archetype, versus the realities of evolution and hence mutability. The concept of a body plan is also somewhat elastic. A few phyla, notably the chaetognaths (arrow worms) and sipunculans (peanut worms), show a remarkable invariance of body plan. One peanut worm looks very much like another. On the other hand, many other phyla show an astonishing range. In the molluscs, for example, the varieties include octopus, garden snail, limpet, and oyster. All these animals evolved, by various routes and pathways, from an original species, the ancestral mollusc. One of the fascinations of zoology is to follow these twists and turns of evolution, tracing out the changes in the key features, such as the muscular foot, that go to define the molluscan body plan. But not only is it possible to recognize a molluscan body plan, but even distinctive body plans within this phylum. Take, for example, the cephalopod molluscs. First, we can construct a cephalopod body plan, and then a whole range of more specific body plans corresponding, for example, to those of the octopus, squid, and extinct ammonite. What these somewhat vernacular and imprecise observations are telling us is that the morphological 'universe' occupied by animals (their morphospace) is not evenly filled, but rather is decidedly clumped at a variety of scales. Hence the molluscan body plan occupies, so to speak, the largest 'cloud', which on closer analysis turns out to be composed of about seven smaller 'clouds' corresponding to major molluscan groups (snails, clams, cephalopods, and four other lesser-known groups). Examined yet more closely then, each of these seven 'clouds' would be seen to be composed of yet smaller clumps, equivalent to the major divisions within, for example, the cephalopods. The inclusion of all these 'clouds' in the Mollusca implies that all are related and by implication the evolutionary transitions between them can be traced with varying degrees of confidence.

What then of the phyla, those 35 basic body plans? We must now imagine 35 'clouds', hanging in morphospace and isolated from one another. Unless these phyla evolved completely independently, which in fact is most unlikely, then somehow we should be able to work out their interrelationships. Take, for example, the phyla known as the molluscs, annelids, and brachiopods. Any zoologist would be able to give you a quick thumbnail sketch of their respective body plans; but what of their evolutionary relationships? Deciding whether molluscs are closer to annelids or brachiopods is highly controversial. The reason for this is that each of these phyla appears so different, so distinct, that it is extraordinarily difficult to imagine how they might have evolved, either from each other or from some other phylum. Any such exercise in imagination usually ends up with the depiction of some sort of generalized worm from which any phylum is derived by a set of more or less arbitrary steps. While this presents a major evolutionary problem, some possible solutions to which are explored below, it paradoxically makes the concept of phylum/body plan strangely attractive. The concept has considerable utility. There may be millions of species, living and extinct, but practically without exception any of them can be conveniently placed in one of the 35 or so phyla. The penalty for such pigeon-holing may be an undue neglect of how phyla actually evolve, but it does help to explain why the Burgess Shale rose to such prominence. This is because not only did this extraordinary deposit house a fair number of species that could readily be housed in familiar phyla, but more significantly there was an apparently weird bestiary with animals quite unlike anything ever seen on the planet.

With the proposal that these strange new animals were representatives of previously unrecognized body plans, however, the magnitude and scope of the Cambrian 'explosion' appeared to have been seriously underestimated. In the opinion of scientists such as Steve Gould, so great was this 'explosion' that it is perhaps futile to invoke any of the normally accepted mechanisms of evolution. This belief, although perfectly understandable, seems to rest on a simple confusion.

What this chapter will try to show is that the strangeness of the problematic Cambrian animals is really a human artefact, a construct of our imagination. This is not to deny that a number of these animals remain very difficult to understand, especially in terms of their phylogenetic relationships. Nevertheless, we can now be confident that ultimately these too will be explained. How can we be so optimistic, when some palaeontologists still argue that the existence of the Burgess Shale problematica threatens to undermine a significant portion of evolutionary theory? Here, I shall introduce two case studies that seem to dispel some of the mystery that has surrounded these peculiar-looking animals. If these examples are accepted, then there is surely good reason to think they have a general applicability to our understanding the Cambrian 'explosion'.

The Burgess Shale arthropods: early views

The history of how the Burgess Shale arthropods were treated is somewhat complex, but it is a story worth understanding. Charles Walcott never had time to assess in detail the relationships of the horde of new arthropods he had uncovered, and he was evidently content to place them in well-known groups, such as the crustaceans. It should also be pointed out that he was primarily a geologist, and a very great one at that. He was not a zoologist, but he certainly relied on the advice of zoologists. For instance, in interpreting various fossils such as sea cucumbers he clearly listened to the opinions of the relevant expert, Austin L. Clark.[1] Similarly, no sooner had he published his description of *Aysheaia*[2] than other zoologists were quick to point out how similar it was to the living *Peripatus*, something that Charles Walcott had overlooked. There is little doubt that had Walcott had the opportunity to restudy *Aysheaia* he would have had no hesitation in accepting this opinion.[3]

How did the Cambridge school, notably Harry Whittington and Derek Briggs, deal with the arthropods? To understand the initial stages of their thinking, it is necessary to introduce one of the greatest figures in arthropod research, Sidnie Manton (1902–79). Her life was largely devoted to painstaking studies of the anatomy of arthropods, especially the insects and the group known as the myriapods (these include the centipedes and millipedes). She was also very interested in the way that arthropods functioned. In particular she produced elegant studies of how arthropods walk and run. One of the main conclusions of her lifetime of scientific endeavour was that the arthropods must have evolved several times.[4] Such multiple origins are known as polyphyletic evolution. In practice, what this means is that a feature (say the jointed leg of a fly, a crab, a trilobite, or a spider) that might be thought to be entirely characteristic of arthropods, could not actually be the same in terms of its evolutionary derivation, that is, arising from a common ancestor, no matter how similar they appeared to be. Sidnie Manton proposed that each major group of arthropods had evolved from some sort of soft-bodied ancestor, the exact nature of which was left rather vague. She thus rejected the idea that all these arthropods had jointed legs because they shared a common ancestor in the late Precambrian. At first sight, this is by no means an unreasonable opinion. Perhaps the jointed legs and other features that are taken to characterize the major groups of arthropods genuinely had evolved independently, in response to a common need? Deciding whether this analysis is correct is certainly not a trivial problem. Again and again when a zoologist looks at a particular feature of an animal, he or she finds it very difficult to decide between the following two options. Is an apparently identical character present in two animals because they share a common ancestor, or is it simply because the number of biological options available to provide a particular function is severely limited?

The guiding principle that Sidnie Manton used in deciding upon arthropod polyphyly was simple, but I believe she was mistaken. Her meticulous studies of arthropod anatomy had revealed a basic and separate identity in each of the major groups. She concluded, therefore, that each group was so distinctive that it was impossible to envisage how a transition between any of them could have occurred. Thus, in her opinion, the arthropods could not be treated as a single phylum. Instead she identified four separate phyla (Fig. 72). These were the chelicerates (including the spiders and scorpions), the crustaceans (animals such as crabs and prawns), the uniramians (principally the insects and myriapods[5]), and the extinct trilobites.

It was with this broad framework in mind that the Cambridge school first embarked on their study. Was the evidence from the Burgess Shale consistent

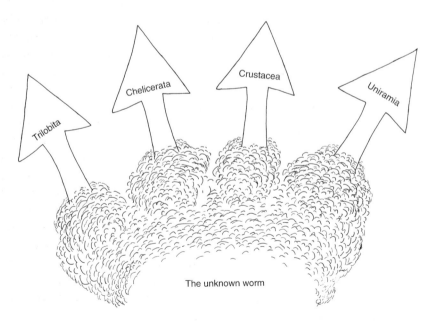

Fig. 72. Arthropod evolution as envisaged by Sidnie Manton, with the four phyla each arising from an unspecified ancestral 'cloud' of forms. The implication of this diagram is that each of the phyla acquired the features typical of arthropods, such as the jointed exoskeleton, independently of the others.

Fig. 73. The Burgess Shale arthropod *Canadaspis perfecta*.

with Sidnie Manton's viewpoint? In the Burgess Shale fauna there are, of course, undoubted trilobites, including *Olenoides* with its beautifully preserved appendages (Fig. 37). Then there are examples of what appear to be crustaceans, most notably an animal known as *Canadaspis* (Fig. 73), and the uniramians, in the shape of *Aysheaia* (Fig. 39). Somewhat later in the research programme, a representative of the chelicerates materialized in the form of *Sanctacaris* (Fig. 74).[6] Thus, representatives of all of Sidnie Manton's supposed phyla were present in the Cambrian. This in itself is not so surprising, because many of other animal phyla were known to appear at about the same time. The problems were not with these few arthropods, but all the others in the Burgess Shale. First, there were a number of fossils that recalled a particular group, but did not seem to be quite right. *Waptia*[7] (Fig. 75) looks like a crustacean, but differs in certain ways. Another arthropod, *Sidneyia* (Fig. 76)[8] has certain features quite characteristic of the chelicerates, but it does not seem to have enough of them to make it a 'true' example. Other arthropods, notably *Marrella* (Fig. 16), seem to resemble trilobites and crustaceans in some respects, but certainly could not be interpreted as genuine examples of either class. Yet others, such as *Odaraia* (Fig. 45), verge on the ridiculous with the fluke-like tail. This particular feature has never been seen in any other arthropod.

In 1979, a time that may reasonably be regarded as the culmination of the first stage of the Burgess Shale investigation, Harry Whittington published a diagram of arthropod evolution as he saw it,[9] basing his conclusions on the

Fig. 74. The arthropod *Sanctacaris uncata*, from the Middle Cambrian of Mount Stephen, British Columbia. [Photograph courtesy of D.H. Collins and the Royal Ontario Museum.]

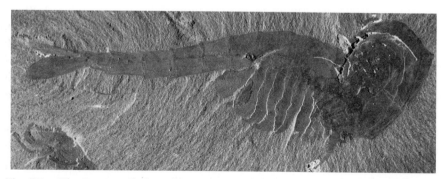

Fig. 75. The Burgess Shale arthropod *Waptia fieldensis*. [Photograph courtesy of C.P. Hughes, formerly University of Cambridge.]

Fig. 76. The Burgess Shale arthropod *Sidneyia inexpectans*. [Photograph courtesy of D.L. Bruton, University of Oslo.]

framework already proposed by Sidnie Manton. Since then this illustration has achieved a certain notoriety, and perhaps even scorn, in some quarters. In fact, in the light of what was then known, it was remarkably honest. A simplified version of this diagram is shown in Fig. 77. Note the four main groups of arthropods, three persisting to the present day, but the trilobites becoming extinct at the end of the Permian, about 250 Ma ago. In between these four groups Harry Whittington added a large number of lines, each one indicating a lineage of arthropods that could not be readily accommodated in any of the supposed phyla: chelicerates, crustaceans, trilobites, and uniramians. This diagram emphasized three things. First, the geological record, and especially the Burgess Shale, was full of arthropods that could not be placed

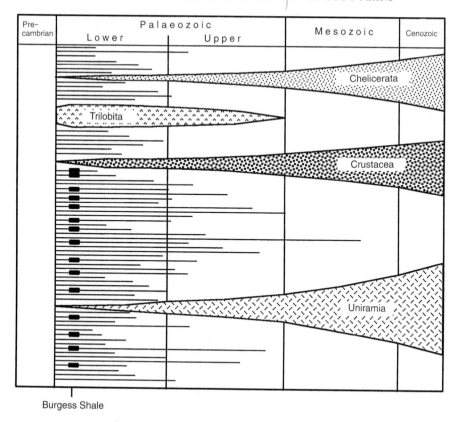

Fig. 77. The evolution of the fossil arthropods as envisaged by Harry Whittington in 1979. [Modified version of fig. 2 of H.B. Whittington (1979). Early arthropods, their appendages and relationships. *In* M.R. House (ed.) *The origin of major invertebrate groups. Systematics Association Special Volume* 12, pp. 253–68. Academic Press Ltd, London.]

in any one of the four major groups. Second, the interrelationships between all these arthropods were very obscure. In particular, Harry Whittington made no attempt to extend these lines into the Precambrian and so link those lineages which he thought to be most closely related. Third, he indicated that many arthropods were yet to be discovered, because those already known were highlighted. Unacknowledged at the time was the presumption that new discoveries might provide 'missing links' between what appear to be very different arthropods. As we shall see below, this appears to be the case.

Cladistics to the rescue?

The general appearance of diagrams of this sort leads them to be dubbed 'phylogenetic lawns'. Why do some scientists regard such depictions with contempt? To explain the background, we need to introduce a method of classification known as cladistics.[10] To many biologists this is regarded as the only correct and proper way to address the problems of evolutionary relationships. The battle to persuade biologists of the central importance of cladistics has been vigorous, often acrimonious, and sometimes even bitter. It is now clear that the cladistic method has won the day. From our perspective, much of the discussion between cladists and those biologists who espoused other systems of zoological classification can be seen to have been rather futile. More curiously, the real problems with cladistics have received rather little attention. But first we need to understand the outlines of cladistics.

The basis of cladistics is very straightforward. In principle it is not open to argument if one accepts the basic precept of evolution that it occurs by descent with modification. In essence any organism can be considered as being composed of a whole series of characters (or, more technically, character-states), that one way or another have been altered from the ancestral state and so are said to be derived. Most obvious would be features of anatomy. These could, for example, be parts of the skeleton, details of soft tissue, or the structure and arrangement of a given organ. But there are also other types of character. For example, there are data from molecular biology. Consider the sequence of amino acids in a protein such as haemoglobin, or alternatively the chain of nucleotides that go to define a gene in a strand of DNA. The extent of the differences between these molecular codes in different animals will provide characters that will help to establish whether the species are distantly or closely related. As mentioned above (p. 151), one of the many reasons for believing that chimpanzees are very close to humans is the identical sequence of amino acids in the blood protein haemoglobin. Even features of an animal's behaviour may be employed. For example, one might want to compare the songs of different species of birds. The recorded sounds can be transformed into spectograms, and their analysis may well reveal characters of taxonomic value. In principle, any or all of these derived characters may be used to define an organism and be used in cladistic analysis.

The fundamental principle of cladistics is that the greater the number of characters (of whatever type) an animal shares with another, then the more closely related they are likely to be. Let us take a simple example, using a type of diagram called a cladogram (Fig. 78). Consider a goldfish, a cat, and a human. It is just conceivable that they evolved from completely different ancestors, but it is much more likely that they share a common ancestor. This is because in all three types of animal there is the key feature of the vertebral column, the precursor of which we met in the notochord of *Pikaia*. This feature tells us nothing about whether we humans are more closely related to cats as against goldfish. In cladistic terminology, which is seldom accused of euphony, the vertebral column is known as a plesiomorphic character. All this means is that if we decide to use this feature alone in our evolutionary analysis we would

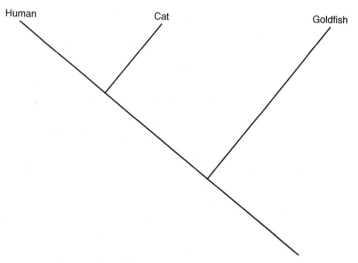

Fig. 78. A very simple cladogram, showing the relationships between a goldfish, a cat, and a human.

not be able to decide whether we humans should greet either cats or goldfish as our nearest cousins. It will be equally apparent, however, that there are many other features, such as hair and the possession of four limbs, that make it far more likely that we are more closely related to cats than we are to goldfish. In fact, the common ancestor of cats and humans lived about 60 Ma ago, whereas the common ancestor of goldfish and humans lived about 500 Ma ago.

So far, so good. What, then, is there to stop us accumulating as much data as possible on as many characters as we can find or measure from all the main groups of animals? With all this information we could then construct the definitive cladogram that would reveal once and for all the evolutionary relationships between all these groups. Not only would we confirm the self-evident (cats and humans are more closely related to each other than either is to the goldfish) but we could answer questions that have puzzled zoologists for hundreds of years. Are the annelids more closely related to the arthropods or to the molluscs? Are the sponges more primitive than the cnidarians? What looks simple in principle is actually very difficult in practice.

First, there is the problem of what is known as evolutionary convergence.[11] Recall the example of the jointed appendage in arthropods. Today, the majority of zoologists accept that this is a plesiomorphic character of arthropods, although as we shall see below it now seems likely that we can investigate the earliest stages of arthropod evolution when they were actually acquiring this feature. Just suppose, however, that Sidnie Manton's ideas are correct and that the jointed appendage has been acquired at least four separate times. If this character is convergent and thus polyphyletic, it can play no part in a discussion of relationships. The fundamental problem here is that we have almost no theoretical understanding of what the constraints and limits of animal 'design' really are. If, as appears to be the case, there are indeed strong constraints, then we must expect convergence to be rampant.[12] At the moment it is still very difficult to decide between characters that are of genuine use in determining evolutionary relationships as against those that have arisen by convergence from unrelated ancestors.

The second problem is that some biologists who wish to embark on a cladistic analysis and the construction of a cladogram are very industrious about assembling a huge database, but too often are less critical in deciding whether the characters chosen are really valid. Suppose, for example, that a whole set of characters are actually interrelated. If one of them changes, so perhaps do most or all of the others. Given that most anatomical characters form functional complexes that operate as integrated units, it may be necessary to focus on only one such character. To make things even worse, we still know far too little about the underlying genetic mechanisms that code for various parts of the anatomy. A shift in the genetic code might conceivably alter several, apparently unrelated, features in the animal.

Much optimism has been expressed about the possibility of molecular data providing an independent set of characters, especially if they have no obvious relationship to the anatomy. This is an important argument, but even molecular information is not free of major problems. Molecular sequences, say of DNA, may be duplicated, reversed, or deleted; or they may change very quickly, so that resemblances to the ancestral state are difficult to detect. In addition, there is also some clear evidence for convergence in molecular sequences. Perhaps the most famous example is that of mammalian lysozymes, whose appearance is linked to the onset of ruminative digestion.[13]

Let us suppose, however, that by one way or another we have a data-set of genuinely reliable characters. Our problems are still not at an end. In principle the construction of a cladogram is simple; in practice it is often enormously difficult. First, the amount of data is usually so large that a computer is essential for processing it. There are, as one might expect, standard programmes available to zoologists. I have been told that there are, however, formidable problems in the application of the underlying mathematical principles, although these are too seldom understood by the zoologists. Even worse, the power of existing computers is far too restricted to deal with anything more than a rather small number of taxa at any one time. In reality even establishing a cladogram for about thirty species based on the comparison of about a hundred characters may take years of continuous computing. For really large data sets the answer the zoologist would like to obtain might take millions of years! To make matters even worse, the computer programme will in many instances give a large number of solutions, all of which are equally plausible.[14] How is the zoologist to decide which one is correct?

Although seldom remarked upon, all this seems rather ironic. As such, cladistics is an unassailable and watertight methodology. In practice, it is haunted by problems at every step, from deciding which characters to employ, how best to process the data, and how to choose the most likely cladogram. These remarks must not be taken out of context. I am not advocating that we abandon cladistics. Advances continue to be made, and the use of molecular versus morphological data holds a real promise of cross-checking. Functional analyses also help to constrain the likelihood of character complexes changing together. The most important fact to remember is that any cladogram is merely a hypothesis, although too often cladograms are presented as being the last word in the debate.[15]

This lengthy discussion is important because the methods of cladistics are now being widely employed in attempting to make sense of the Cambrian 'explosion', and especially the position of the so-called problematic taxa. We need to turn, therefore, first of all to the problem of the Cambrian arthropods. Earlier in this discussion I introduced the concept of a phylogenetic 'lawn' (Fig. 77). Such diagrams have an obvious disadvantage that once presented there is not very much more you can do with them. They are in effect statements of ignorance. This is because they fail to formulate any hypothesis of relationship that can be tested, either by the acquisition of new data or by reconsideration of existing information. With the enthusiasm for cladistic analyses sweeping all areas of biology, it was hardly surprising that the arthropods of the Burgess Shale did not escape the net of enquiry.[16]

Until now we have seen a steady improvement in our understanding of the phylogeny of these arthropods. It is important, of course, not to consider the Burgess Shale examples in a vacuum, out of a context that can include other arthropods, Accordingly the cladograms also need to incorporate a variety of other arthropods, both living and fossil. The most revealing feature of one of the first attempts[17] at a cladistic analysis was the overall configuration of the cladogram (Fig. 79). It looked somewhat like a comb, and in language of the cladistic adept it is referred to as pectinate cladogram. Its comb-like appearance arises because, with few exceptions, each taxon lies at the end of branch, with a succession of such branches making up the cladogram. This pectinate arrangement can be contrasted to many other cladograms that tend to have a more bush-like appearance.

The first of the cladograms to incorporate the Burgess Shale arthropods considered a relatively large number of taxa and different morphological characters. To construct this cladogram a standard computer programme was employed. Nevertheless, the general robustness of the cladogram, measured by using a fairly simple statistic, was rather low. This measurement implied that however interesting this cladogram might be as an hypothesis of arthropod relationships, it certainly could not be regarded as very secure or reliable. The pectinate arrangement of the cladogram has another important implica-

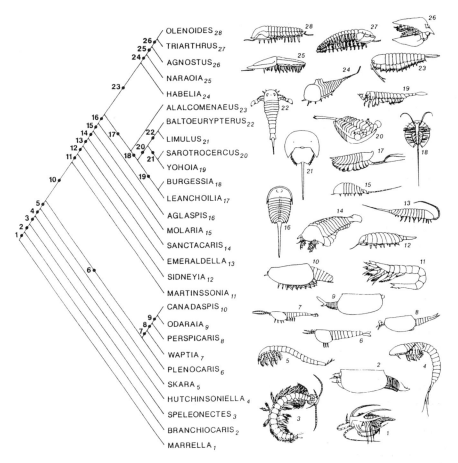

Fig. 79. An early attempt at a cladogram of the arthropods, with an emphasis on the Burgess Shale representatives. [Reproduced with permission from fig. 1 in R.A. Fortey (1990). Trilobite evolution and systematics. *In* D.G. Mikulic (ed.) *Arthropod paleobiology*, pp. 44–65. Short Courses in Paleontology 3. Paleontological Society.]

tion. The arrangement of the branches is in effect a series of steps, each one being defined by a few, or perhaps only one, change in character. It should be apparent that even quite minor changes in either our understanding or our knowledge of these morphological characters might therefore make a substantial difference to the order of branching. Nevertheless, despite its potential sensitivity to such changes, this preliminary cladogram revealed some quite interesting features. Perhaps the most unexpected feature was that trilobites, almost universally considered to be rather primitive arthropods, actually transpired in this analysis to be advanced and also quite close to the chelicerates.

Since then we have seen a progressive refinement in the cladograms. Interestingly, the latest versions[18] have lost their pectinate configuration and

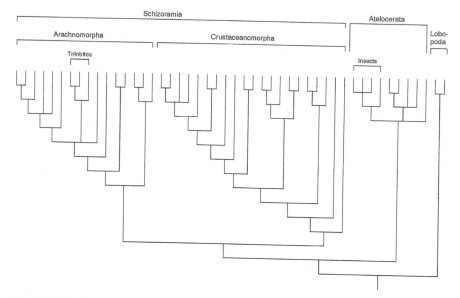

Fig. 80. A more recent cladistic analysis of the arthropods. [Redrawn from fig. 11 of M.A. Wills, D.E.G. Briggs, and R.A. Fortey (1994). Disparity as an evolutionary index: a comparison of Cambrian and Recent arthropods. *Paleobiology* 20, pp. 93–130. Paleontological Society]

are now much more bush-like in shape (Fig. 80). We can now make two cardinal observations. First, despite a monophyletic origin for the arthropods, three main groups remain distinguishable. These are referred to as the Arachnomorpha, Crustaceanomorpha, and Atelocerata, and represent expanded versions respectively of the chelicerates (including now the trilobites), the crustaceans, and the insects and their near-relatives. Even these proposals, however, cannot be regarded as watertight. For example, evidence from molecular biology now indicates that the crustaceans and insects are much more closely related than was once thought.[19] Second, and of equal importance, is that the various Burgess Shale arthropods nest within these groupings. If these species really were as peculiar and bizarre as some palaeontologists have proposed, then even in the revised cladograms they would stick out as isolated branches. Take, for example, the arthropod *Odaraia* (Fig. 45). Notwithstanding the extraordinary and unique posterior tail-fluke, the general characters of this animal place it firmly within the Crustaceanomorpha.

In the most recent cladograms the most primitive group appears to be the lobopodians, of which the best understood example is *Aysheaia* (Fig. 39). As we saw above there is, however, abundant evidence that the Cambrian lobopodians were in the throes of a very substantial radiation. The most notable example, at least in the eyes of the public, is *Hallucigenia* (Figs 18, 19), but there is also a bevy of related animals from Chengjiang[20] and an important discovery from the Baltic region in the form of a fossil known as *Xenusion*.[21] Ultimately these various lobopodians will need to be built into the next generation of cladograms, even though *Aysheaia* may continue to serve as a convenient proxy. But should we stop with the lobopodians? Could we not also plan to incorporate animals, all extinct and so only known from fossils, that in a sense form stepping stones towards the phylum of Arthropoda from a worm-like ancestor of some sort? In other words, these would be species that had acquired some, but—most important to understand—not all, of the features that define the complete arthropod. In the previous sentence the words 'not all' are especially important.

The resistance to such ideas is quite remarkable, even though in principle it is entirely consistent with our general understanding of evolutionary processes. Why should this be? I believe that the underlying reasons are twofold. First, humans have an innate desire to classify and so pigeon-hole their concepts. When does an arthropod become a true arthropod? No palaeontologist would deny that the trilobites or *Marrella* are genuine arthropods, but as we shall see below some are distinctly unhappy with the inclusion of *Anomalocaris*. So what do they do? They make a new pigeonhole and say: '*Anomalocaris* must belong to a new phylum.' But this is really an evasion and solves nothing, at least in the context of evolution. Second, there are persistent claims that the origin of phyla such as the Arthopoda or supposed extinct phyla can be explained only by new mechanisms of evolution.[22] However bold and exciting this claim appears to be, I suspect that it is without foundation.

Cambrian arthropods: the emergence of a synthesis

So how might we unravel the steps in evolution that led to the true arthropods? It will unfortunately be some years before the ideas that are beginning to emerge can be tested rigorously. Indeed, we are still at a preliminary stage. First, let us consider *Anomalocaris*. Its morphology hints at some of the complexities in our understanding of the very early evolution of the arthropods. Its one obvious arthropod character is the pair of jointed appendages attached to the head. There is, however, published information showing that close relatives of *Anomalocaris* had a series of jointed legs.[23] Unpublished work[24] on other anomalocarid-like animals indicates that, instead of jointed legs, the appendages were represented by lobopods quite similar to those of *Aysheaia*. If this is correct, then *Anomalocaris* probably walked over the sea floor, rather than, as had been thought, swimming with the flap-like extensions.[25] From being considered as one of the most enigmatic of Cambrian animals, a true evolutionary conundrum, *Anomalocaris* (and some closely related species) is now beginning to be regarded as belonging to a group of very primitive arthropods. At the moment, it appears that the lobopodians, which include *Aysheaia* and *Hallucigenia*, are even more primitive. If the presence of lobopod-like limbs in *Anomalocaris* is confirmed, then its flap-like extensions above these legs may be a new feature. These flaps may have combined several functions: protection, propulsive organs, and respiratory surfaces. Of these, perhaps the last was the most important. Imagine that in due course the flaps seen in *Anomalocaris* were modified. The leading edge now formed an elongate bar and behind it the structure was transformed in a series of trailing filaments. This, of course, would be reminiscent of the gills that arise above the walking legs of many arthropods. Thus, if the lobopod limbs were transformed into jointed appendages and the flaps into gills, then one could envisage, at least in broad outline, the transformation between an animal similar to *Anomalocaris* and a fully fledged arthropod.[26]

What about the complex mouth apparatus, apparently unique to *Anomalocaris* and its relatives? One possibility is that it was ultimately lost, and the appendages adjacent to the mouth took over this function by becoming specialized in the handling, shredding, and passing of food.[27] All this is very speculative indeed, and really raises more questions than it answers. The central fact remains, however, that *Anomalocaris* is not a member of some weird new phylum but appears to be a fossil crucial to our understanding of the early stages in the emergence of arthropods. In addition, it needs to be stressed that no matter how primitive it is as an animal, in the context of the Cambrian it was a highly effective predator.

There is another suggestion, made by Graham Budd, that *Opabinia* also possessed a set of walking lobopods.[28] It now seems likely that *Anomalocaris* and *Opabinia* are quite closely related. A particularly striking feature present in both *Anomalocaris* and *Opabinia* is a prominent tail-fan. That of *Opabinia* has been long recognized, but its presence in *Anomalocaris* was more recently established on the basis of fossils collected by Des Collins of the Royal Ontario Museum.[29] At some point in the future we shall also have to consider in detail other apparently strange animals, most notably *Kerygmachela* from the Sirius Passet fauna[30] (Fig. 81). In this fossil there is also good evidence both for lobopod limbs and for the flap-like extensions of the body, recalling what is seen in *Anomalocaris* and *Opabinia*. Exactly how the giant grasping apparatus, at the front of *Kerygmachela*, can be compared with the anterior structures in the *Anomalocaris* and *Opabinia* is still to be determined.

Let us accept that all these animals will ultimately tell us much about the early evolution of arthropods. How much further back shall we be able to trace the roots of the arthropod tree? This is still very uncertain. Some clues might come from Ediacaran animals such as *Spriggina* (Fig. 82) and a more

Fig. 81. The Sirius Passet arthropod *Kerygmachela kierkegaardi*.

poorly known fossil named *Bomakellia*.[31] Exactly how these relationships might emerge, nevertheless, is difficult to judge. Comparisons between *Spriggina* and the arthropods have a long history, especially because the Ediacaran animal has an anterior end that seems to consist of a prominent head-shield, vaguely reminiscent of the trilobites.[32] Ultimately, *Bomakellia* may turn out to be a more promising candidate, especially because the animal bears plate-like extensions that can perhaps be compared to the lobes of *Anomalocaris*.

The story of *Wiwaxia*

It is generally assumed that the arthropod lineage will be traced back to some sort of flatworm. The arthropods undoubtedly form a major phylum within the protostomes, but their relationships to the other protostome phyla are not very clear. Evidence from molecular biology suggests that the arthropods may have branched off rather early[33] and are somewhat isolated from another

Fig. 82. The Ediacaran animal *Spriggina floundersi*, which many palaeontologists believe may be related in some way to the arthropods.

major cluster that encompasses a number of phyla, including the molluscs, the annelids, and the lophophorates (best known in the fossil record from the brachiopods, but including also two other phyla: the bryozoans and phoronids). Does the fossil record reveal anything about the steps that led to any of these phyla? Not long ago such a claim would have been regarded as rather far-fetched. Research now suggests that just as with the arthropods we may be seeing the glimmerings of an answer. For this part of the story we need to return to two key fossils: *Wiwaxia* from the Burgess Shale (Fig. 43; Plate 3) and the halkieriids from Sirius Passet (Figs 53, 67).

First, let us reconsider *Wiwaxia* (Fig. 17). The present position is easier to understand if we first review the history of the investigation. Charles Walcott considered this animal to be a polychaete annelid.[34] He bolstered this conclusion by comparing the sclerites (the plate-like skeletal units that, although individually inserted via a short stalk into the body of *Wiwaxia*, together formed a defensive coat as a cataphract armour) to scale-like structures (known as the elytra) that cover the upper side of certain polychaetes, such as a group referred to as the polynoids. When I came to restudy the specimens, I was more impressed with the similarities between *Wiwaxia* and what we believe primitive molluscs to have looked like. The reasons for reaching this conclusion were as follows. The undersurface of *Wiwaxia* lacked sclerites, but instead formed what appears to have been a soft sole. This structure appears to be very similar to the creeping muscular foot of some molluscs. It is best seen in groups such as the snails (think of the slug), as well as more primitive types that include the monoplacophorans. This latter group flourished in the Lower Palaeozoic and was discovered in the 1950s to be living in deep oceans in the form of the 'living fossil' *Neopilina*.[35] In what many believe to be an even more primitive group, known as the aplacophorans,[36] the foot is still present, but is reduced in size. In addition, *Wiwaxia* possessed a jaw that looks quite similar to the feeding apparatus of the molluscs, where it is known as the radula.

It is generally accepted that the aplacophorans, a rather obscure group well understood only by a handful of specialists, are our best glimpse of the likely appearance of the most primitive molluscs.[37] These animals look like some sort of spiny worm because, in contrast to other molluscs, which have a shell of some sort, in the aplacophorans the surface is coated with numerous calcareous spicules. Many zoologists believe that the shells of the other molluscs originated by the fusion of these spicules. It must be admitted that there are obvious differences between the sclerites of *Wiwaxia* and the spicules of aplacophorans, but the general appearance of the Burgess Shale animal seemed to be consistent with a position fairly close to the earliest molluscs.

The scientific tables then appeared to be turned back in favour of *Wiwaxia* being a polychaete annelid, thanks to remarkable work by Nick Butterfield.[38] By the utmost delicacy of preparation in the laboratory he managed to isolate individual sclerites of *Wiwaxia*. The specimens he freed from the surrounding shale matrix were very small, and presumably came from juvenile specimens that themselves were only a few millimetres in length. When Nick Butterfield looked at these specimens under high-power magnification, he saw that the walls of the sclerites were translucent and, more importantly, had a very characteristic microstructure. This consisted of a closely parallel lineation. More surprises were to follow. Using the same techniques of acid digestion he was able to extract examples of the chaetae that in life were inserted into the body of the Burgess Shale polychaete annelid *Canadia* (Fig. 44). They too showed the same type of microstructure, which not surprisingly was closely similar to that seen in the chaetae of living polychaetes. Recall how the chaetae are arranged in a typical polychaete, such as we see in the Burgess Shale *Canadia* (Fig. 44). In this animal, which seems to be particularly primitive, each segment of the body bears on either side two separate bundles of chaetae. The upper bundle, the notochaetae, arises from the dorsal surface and provides a roof of protective spines. The lower bundle forms the neurochaetae. They arise from a lobe-like structure and are responsible for locomotion. Each bundle acts as a sort of leg that can push against the sea floor and so lever the worm forwards.

Nick Butterfield therefore concluded that Charles Walcott had been right all along. *Wiwaxia* must be a genuine polychaete. But what about the evidence that *Wiwaxia* had important similarities with the molluscs? Was one interpretation simply incorrect, or could these differences be reconciled? At about the same time that this work was being carried out, researchers in the area of molecular biology were providing important new insights into metazoan phylogeny on the basis of their work into molecular sequences, notably the molecule known as ribosomal RNA. One thing that emerged was that molluscs and annelids were much more closely related than had hitherto been generally accepted.[39] Indeed, until then, most zoologists had believed annelids to be closely related to the arthropods, while they regarded the molluscs as some sort of off-shoot of the flatworms. The proposals from the molecular biologists were not in fact exactly new, but such ideas had become distinctly unfashionable.

Accepting that the molluscs and annelids are phylogenetically closer than had been generally thought is an exciting conclusion, but it is only a first step. Most importantly, it leaves unsolved the problem of what the common ancestor of these two phyla looked like. Taking even the most primitive mollusc, by general consensus an aplacophoran, and the most archaic annelid, probably a marine polychaete, it is far from clear how one might have evolved from the other. They simply look too different.

News from Greenland

To begin to see how we might supply an answer to this difficulty, we need to turn to the halkieriids. This group is long gone, totally extinct, so our only information is from the fossils (Fig. 53). If one were brought back to life most probably it would be described as an armoured slug (Fig. 67). What this means is that the undersurface of the animal was soft, but muscular enough to enable the halkieriid to glide over the sediment. The upper surface of the animal, however, was not protected by a rubbery skin, but was covered by a cataphract-like armour of sclerites, totalling about 2000 in an adult individual, as well as a large shell at either end of the body. At first sight halkieriids (Figs 53, 67) and *Wiwaxia* (Fig. 43) appear to be only superficially similar.[40] To be sure, both animals bear sclerites, but there one might think the resemblance stopped. Closer examination, however, suggests that they are in fact quite closely related. First, halkieriids and *Wiwaxia* appear to have the same basic arrangement of sclerites, divisible into three broad zones. Thus, the sides of the animal are mantled by lateral (that is cultrate) sclerites. This distinctive zone separates the dorsal region with its palmate sclerites from the ventro-lateral division with its curved siculate sclerites. The sclerites in both animals are generally similar in shape and ornamentation. They are hollow[41] and have a basal opening through which extended a stalk that was embedded in the body wall. The sclerites of halkieriids are calcareous, whereas those of

Wiwaxia are unmineralized. This difference is probably less important than at first might be thought. In many other animal groups the occurrence of bio-mineralization is sporadic, only some species having mineralized hard parts. There are two other obvious differences. One is trivial, the other of greater significance. First, in contrast to the Greenland halkieriid the Burgess Shale *Wiwaxia* has elongate spines that project above the rest of the body. Despite their dramatic appearance these spines are probably only cultrate sclerites modified by elongation. In terms of phylogeny this difference is very minor, but this may not be the case in terms of ecology. Lacking mineralized sclerites, the prominent spines of *Wiwaxia* may have helped to confer additional protection. It may also have been the case that the intensity of predation had escalated from the time of the Sirius Passet fauna to that of the Burgess Shale (Fig. 56). The other difference is almost self-evident because the Greenland halkieriid carries two prominent shells, whereas such structures were either greatly reduced or absent in *Wiwaxia*.[42]

Let us accept that the halkeriids and *Wiwaxia* are indeed related. How can we fit them into any sensible scheme of metazoan phylogeny? Figure 83 shows the outlines of one possibility. The first point is that Nick Butterfield was quite correct to compare the sclerites of *Wiwaxia* to the chaetae of polychaetes, but he was incorrect to conclude on the basis of this feature that *Wiwaxia* must belong to this group. How is it that *Wiwaxia* possesses a diagnostic feature of the polychaetes, but cannot be accepted as a true member of

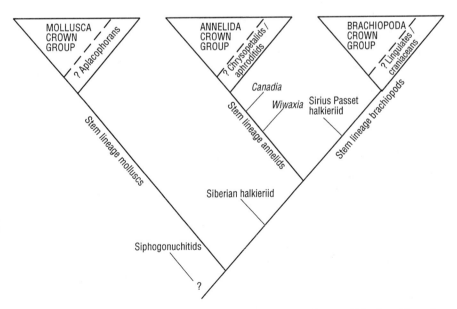

Fig. 83. The possible evolutionary relationships between the annelids, brachiopods, and molluscs, with the proposed position of the halkieriids and *Wiwaxia*.

the group? The answer appears to be that *Wiwaxia* is on the path of evolution that will lead to the polychaetes, but it has not gone all the way.

So the transition from halkieriid to wiwaxiid, and thus ultimately a polychaete annelid, may have proceeded as follows (Fig. 84). In the Greenland halkieriids the siculate sclerites form a long series of closely imbricated bundles that run along either edge of the body. In detail, each bundle of sclerites appears to have arisen from a lobe-like extension of the body. A somewhat similar arrangement appears also in *Wiwaxia*, except that in this animal the total number of bundles of siculate sclerites is much reduced. In addition, its sclerites are considerably larger. I would suggest that these siculate sclerites are ancestral to the bundles of neurochaetae that we find in the polychaetes. It seems rather unlikely, however, that the siculates of either halkieriids or *Wiwaxia* could have served to propel the worms across the seabed as they do in the polychaetes. First, they are closely packed together and do not appear to have had the freedom of movement that would be necessary if they were to perform the stepping motion that is executed by the bundles of neurochaetae. Second, in both groups the sclerites are strongly curved and appear to be of a rather unsuitable shape to push against the sea floor during locomotion. If they could not walk, then how did halkieriids and wiwaxiids progress across the seabed? Presumably these worms moved by muscular contractions running across the ventral surface, in a manner similar to a slug. The transition from such a primitive gliding motion, which is still employed by many molluscs, to the stepping locomotion of polychaetes was probably a gradual process. The key innovation in this shift of locomotory style may have lain in the anatomy of the lobes that first bore siculate sclerites and subsequently were transformed into the neurochaetae. The mobility of these structures would have been much enhanced if a fluid-filled internal cavity developed. This would provide an antagonistic system to oppose the surrounding muscles in the body wall. This certainly seems plausible when we look at living polychaetes. The movement of the neurochaetae is closely controlled by the hydrostatic protrusion of the lobe and the associated muscles that impart the locomotory force.

The transition between palmate sclerites and notochaetae is even easier to envisage (Fig. 84). This is because the fundamental function, that of

Fig. 84. Proposed anatomical transitions between the halkieriids (upper, as represented by *Halkieria evangelista*), the wiwaxiids (middle, as represented by *Wiwaxia corrugata*), and a primitive polychaete annelid, *Canadia spinosa* (lower). The reconstructions represent block diagrams of the right-hand side of part of the body, to show both the arrangement of the sclerites (halkieriids and wiwaxiids) and chaetae (polychaete) and a cut-away to reveal the intestine but note that the internal anatomy was considerably more complex than is depicted here). To the right is a detail of the arrangement of the siculate sclerites in halkieriids and wiwaxiids, and the equivalent neurochaetae in the polychaete annelid.

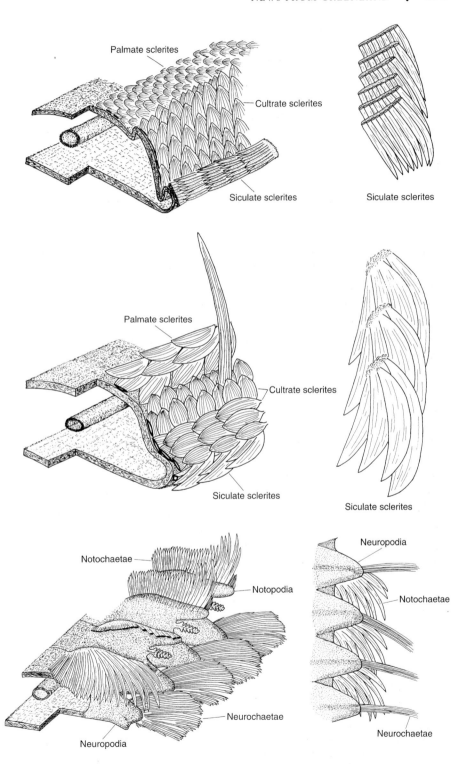

Palmate sclerites

Cultrate sclerites

Siculate sclerites

Siculate sclerites

Palmate sclerites

Cultrate sclerites

Siculate sclerites

Siculate sclerites

Notochaetae

Notopodia

Neuropodia

Notochaetae

Neurochaetae

Neuropodia

Neurochaetae

providing a protective coat, remains effectively unchanged. So if the siculate sclerites transform into neurochaetae and similarly the palmates into notochaetae, what happens to the intervening cultrate sclerites? It seems likely that they experienced successive reduction and ultimately were lost. The main reason for this may be that as distinct lobes developed in association with the notochaetae and neurochaetae, so they defined an intervening and protected recess that made the presence of the cultrate sclerites unnecessary.

Not only does molecular biology indicate a connection between annelids and molluscs, but also a close relationship with the lophophorates (brachiopods and the related phyla of bryozoans and phoronids).[43] This has come as a considerable shock to many zoologists. Why should this be? The reason for their surprise is that zoological tradition has strongly maintained that brachiopods and their relatives are much closer to such phyla as the chordates (which includes the vertebrates) and the echinoderms. Chordates and echinoderms are generally agreed to belong to a major grouping known as the deuterostomes (a term meaning 'second month'; see Glossary). The evidence that zoologists have available certainly made this idea seem very reasonable. First, the embryology of the lophophorates is typical of other deuterostomes. Second, certain deuterostomes (known as the hemichordates) have a tentacular feeding structure that is remarkably similar to the lophophore of the brachiopods and their relatives.[44] It seemed, therefore, to be a cut-and-dried case—that is, until the molecular biologists and palaeontologists joined the debate.

Let us return to the Greenland halkieriid. Its most striking feature is the prominent shell at either end of the body. The posterior shell (Fig. 85) is strikingly similar to the valves of many Cambrian brachiopods, but at first sight this would seem to be simply a rather strange coincidence. But such a judgement might be premature. There appear to be three cardinal facts. First, from the margin of each valve of the brachiopod shell chitinous bristles extend outwards into the surrounding sea water. These bristles, known as the setae,[45] are believed to have a sensory function. It has long been recognized that the microstructure of these setae is identical to the chitinous bristles of polychaetes,[46] which in these animals form the bundles of notochaetae and neurochaetae. As we have already seen, these chaetae can be shown to have evolved from the sclerites of halkieriids. Second, the setae of brachiopods develop early in their embryology, and in some species they have a segmental arrangement. (Recall that the sclerites of halkieriids and the chaetae of polychaetes are also arranged in segments.) Third, in the early stages of development of what may be an especially primitive, but living, brachiopod (*Neocrania*) the animal is first of all mobile and crawls across the substrate.[47] It bears the usual bundles of setae, and although it has not yet started to secrete the two valves of its shell the areas where they will be formed are

Fig. 85. The posterior shell of the halkieriid *Halkieria evangelista* from the Sirius Passet fauna of North Greenland.

already visible. All adult brachiopods are more or less sessile, usually fixed to the sea floor. As part of this transformation from crawling to becoming sessile slightly later in its development the *Neocrania* animal folds about its middle. The net result of this is that the area that will shortly start to secrete the dorsal valve overlies the future position of the ventral valve.

It is now possible to see what may have originally happened back in the Cambrian in order to produce the first brachiopod (Fig. 86). First, imagine a juvenile halkieriid, equipped with its two shells. Because the animal is very small the shells are juxtaposed, back to back, although as it continues to grow they will become progressively separated. At this early stage, however, suppose that the animal developed the ability to swing one shell beneath the other, perhaps as a defensive reaction when threatened. Could this represent the first step towards becoming a brachiopod? Remember also that the

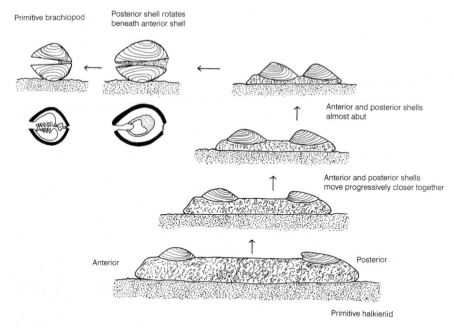

Primitive brachiopod Posterior shell rotates beneath anterior shell

Anterior and posterior shells almost abut

Anterior and posterior shells move progressively closer together

Anterior Posterior

Primitive halkieriid

Fig. 86. The possible transition between a halkieriid and a brachiopod.

halkieriid shells are surrounded by sclerites. Let us accept that these sclerites could transform into the chaetae of a polychaete. If we also accept that these chaetae have a microstructure identical with the setae of brachiopods, then it does not seem fantastic to suggest that the setae of the brachiopods evolved from the sclerites of the halkieriid.

The original events must, of course, have been more complicated. To transform a halkieriid into a brachiopod requires much more than bringing the two shells together. Most important would be the development of a tentacular lophophore, necessary for suspension-feeding as the animal transferred from an active crawling life to a sedentary position on the sea floor. The details of how this might have happened need to be investigated. An instructive parallel might, however, be drawn with the polychaete worms. As we saw above, there is good evidence that they evolved from within the halkieriid group, but remained as active crawlers using the newly evolved neurochaetae for walking across the sea floor. Later in their evolutionary history, however, some of these polychaetes became sedentary.[48] To protect themselves they had to utilize hard parts, and despite having evolved from completely soft-bodied ancestors these worms typically secrete a calcareous tube (or build one by agglutination of sand grains) into which they can retreat in times of danger. The active polychaetes are equipped with jaws, but once a species became sedentary these were replaced by a crown of feeding tentacles. These evolved quite separately from the lophophore of brachiopods and their relatives, but this parallel story shows how the basic steps might have been achieved.

Notes on Chapter 7

1. The type of fossil in question was *Eldonia* and is discussed by C.D. Walcott in *Smithsonian Miscellaneous Collections* (Vol. 57, pp. 41–68 [1911]).
2. Walcott's account of *Aysheaia* appeared in the paper devoted to the worms (*Smithsonian Miscellaneous Collections*, Vol. 57, pp. 109–44 [1911]), because he interpreted it as a polychaete annelid.
3. The correspondence was published in Walcott's posthumous paper on the Burgess Shale, published in *Smithsonian Miscellaneous Collections* (Vol. 85, pp. 1–46 [1931]).
4. S.M. Manton's synthesis of a lifetime's work may be found in her book *The Arthropoda: habits, functional morphology, and evolution* (Clarendon Press, Oxford, 1977).
5. This is not an area of controversy that need detain us here, but there is growing evidence, from molecular biology, that the insects are in fact closely related to the crustaceans. Key papers in this regard are by M. Friedrich and D. Tautz (*Nature*, Vol. 376, pp. 165–7 [1995]) and M. Averof and S.M. Cohen (*Nature*, Vol. 385, pp. 627–30 [1997]). The position of the myriapods (millipedes and centipedes) remains considerably more uncertain, but it is less likely they are close to the insects, and they may be derived from a relatively primitive stock of arthropods.
6. This arthropod was described by D.E.G. Briggs and D. Collins in *Palaeontology* (Vol. 31, pp. 779–98 [1988]).
7. *Waptia* has never been systematically described since Walcott's publications in *Smithsonian Miscellaneous Collections* (Vol. 57, pp. 145–228 [1912] and Vol. 85, pp. 1–46 [1931]).
8. The redescription of *Sidneyia* by D.L. Bruton may be found in *Philosophical Transactions of the Royal Society of London* B (Vol. 295, pp. 619–56 [1981]).
9. This summary of arthropod evolution in the context of the Burgess Shale by H.B. Whittington was published in *The origin of major invertebrate groups* (ed. M.R. House), pp. 253–68. Systematics Association Special Volume 12. (Academic Press, London, 1979).
10. Further details of the cladistic methodology may be found in several sources. These include: E.O. Wiley, D. Siegel-Causey, A.D.R. Brooks, and V.A. Funk. *The compleat cladist: a primer of phylogenetic procedures.* University of Kansas Museum of Natural History, Special Publication 19, i–x, 1–158 (Lawrence, Kansas, 1991); P.L. Forey, C.J. Humphries, I.J. Kitching, R.W. Scotland, D.J. Siebert, and D.M. Williams. *Cladistics: a practical course in systematics.* Systematics Association Special Publication 10, i–xii, 1–191 (Clarendon Press, Oxford, 1992); and A.B. Smith *Systematics and the fossil record: documenting evolutionary patterns* (Blackwell Scientific Publications, 1994).
11. For an excellent review of examples amongst the invertebrates, see the paper by J. Moore and P. Willmer in *Biological Reviews of the Cambridge Philosophical Society* (Vol. 72, pp. 1–60 [1997]).
12. A perusal of the biological literature reveals that convergence is the norm. It is also becoming clear that complexity of organization is no bar to its recurrent appearance, and cannot in itself be used *a priori* as a reliable monophyletic character. See, for example, the comments by A.D. Yoder in *Trends in Ecology and Evolution* (Vol. 12, pp. 86–8 [1997]). Convergence raises many interesting problems on how life is constrained, and equally importantly on the extent to which the expressed architecture of life shares a common genetic basis.

13. A key paper is by K.W. Swanson and others in *Journal of Molecular Evolution* (Vol. 33, pp. 418–25 [1991]). See also the more recent article by W. Messier and C.B. Stewart in *Nature* (Vol. 385, pp. 151–4 [1997]).

14. One of the best examples of this is the highly publicized discussions of the so-called 'mitochondrial Eve', that is, the estimates of the divergence times of modern humans on the basis of comparisons of sequences obtained from mito-chondrial DNA. Fresh analysis of the data quickly produced not only equally parsimonious trees, but even shorter ones than had been first identified. The story of the search for the 'mitochondrial Eve' is nicely summarized by C. Stringer and R. McKie in *African exodus: the origins of modern humanity* (Jonathan Cape, London, 1996). Another example that deserves to be better known is the analysis by S.J. Suter *(Biological Journal of the Linnean Society*, Vol. 53, pp. 31–72 [1994]), which comprised a cladistic analysis of the cassiduloid echinoids. He found more than 70,000 (!) equally parsimonious trees, and reasonably con-cluded that homoplasy was rampant. Subsequently M. Wilkinson and others *(Historical Biology*, Vol. 12, pp. 63–73 [1996]) attempted a 'rescue' operation of Suter's data using another method of cladistic analysis, but this too foundered under the weight of homoplasy.

15. The mantra of terms that surrounds cladistics—maximum likelihood, branch and bound, strict consensus, consistency—gives the flavour of the cladistic enterprise.

16. For the first trial see D.E.G. Briggs and H.B. Whittington in *United States Geological Survey Open File Report*, Vol. 81–743 (Short Papers for the Second International Symposium on the Cambrian System), pp. 38–41 [1981].

17. This paper by D.E.G. Briggs and R.A. Fortey was published in *Science* (Vol. 246, pp. 241–3 [1989]).

18. See, for example, the analysis by M.A. Wills and others in *Paleobiology* (Vol. 20, pp. 93–130 [1994]).

19. See note 4 of this chapter.

20. See notes 27, 28, and 17 in Chapters 3, 4, and 5 respectively.

21. A reconsideration of *Xenusion* was published by J. Dzik and G. Krumbiegel in *Lethaia* (Vol. 22, pp. 169–82 [1989]).

22. This is a complex area, and one upon which a wide variety of opinions has been expressed. Perhaps a fair summary appears in the concluding section (p. 100) of a paper by J.W. Valentine and D.H. Erwin (in *Development as an Evolutionary Process* (ed. R.A. Raff and E.C. Raff), pp. 71–107 (Liss, New York, 1987)), where they write: 'we envision an evolutionary process not unlike forms of selection that operate during microevolution, but with mechanisms of genome change that do not operate at the same intensity or with the same results today'. These authors make it clear in the following sentence that 'these postulated processes do operate at ... the level of natural selection in populations'. Others, however, have pursued a bolder agenda. Thus, in *Wonderful life* S.J. Gould writes (p. 230), 'Perhaps modern organ-isms could not spawn a rapid array of fundamentally new designs,' although he also makes clear that the answer to this question will probably become clear from work in the rapidly moving fields of genetics and developmental biology. And indeed this is happening, with the realization that the genetic mechanisms of animals have a fundamental identity, which may suggest that the apparent gulfs between body plans result from little more than 'tinkering' with the genome.

23. The paper in which these appendages are described is by X-G. Hou *et al.* in *GFF* (Vol. 117, pp. 163–83 [1995]).

24. G.E. Budd, personal communication. The context in which these new observa-tions will be placed is given in his paper in *Lethaia* (Vol. 29, pp. 1–14 [1996]).

25. It is obviously a matter of considerable controversy whether all species of *Anomalocaris* (and closely related genera) were equipped with either lobopods or jointed walking legs, in as much as both taxa from the Burgess Shale (D.H. Collins in *Journal of Paleontology*, Vol. 70, pp. 280–93 [1996]) and

Chengjiang (J-Y. Chen and others in *Science*, Vol. 264, pp. 1304–8 [1994]) appear to lack such structures. The complete absence of such appendages might be surprising, and tentatively it may be better to look towards a taphonomic explanation: the levels at which the shale splits could be controlled by the films that go to make up the anomalocarid body.

26. The detailed arguments for such a transformation are given by G.E. Budd; see note 24.

27. The juxtaposition of the tardigrades to *Opabinia* and *Anomalocaris* by G.E. Budd (*Lethaia*, Vol. 29, pp. 1–14 [1996]; see fig. 9) suggests, however, alternatives in terms of mouth-part homologies which are currently being explored by Graham Budd (personal communication).

28. See note 24.

29. See note 25.

30. The preliminary description of this remarkable animal is given by G.E. Budd in *Nature* (Vol. 364, pp. 709–11 [1993]).

31. It is regrettable that no detailed assessment of *Spriggina*, perhaps the icon of Ediacaran faunas, has been published. The book by M.F. Glaessner entitled *The dawn of animal life: a biohistorical study* (Cambridge University Press, 1984) provides the best introduction to the literature on *Spriggina*, although I am unable to accept Glaessner's detailed comparisons with the polychaete annelids. In a review paper on metazoan phylogeny in *Development, 1994 Supplement* (pp. 1–13 [1994]) I emphasized the possible importance of these and other Ediacaran ?arthropods, and this theme has been more extensively developed by B.M. Waggoner in *Systematic Biology* (Vol. 45, pp. 190–222 [1996]).

32. It has to be said, however, that the reconstructions of *Spriggina* offered by S.J.R. Birket-Smith (*Zoologisches Jahrbuch für Anatomie*, Vol. 105, pp. 237–58) verge on the fanciful.

33. A key paper in this regard was J. Lake's analysis of molecular data in the context of metazoan evolution in *Proceedings of the National Academy of Sciences*, USA (Vol. 87, pp. 763–6 [1990]).

34. Walcott's description and brief discussion of the phylogenetic position of *Wiwaxia* can be found in *Smithsonian Miscellaneous Collections* (Vol. 57, pp. 109–44 [1911]).

35. The classic descriptions of *Neopilina* are by H. Lemche and K.G. Wingstrand (*Galathea Reports*, Vol. 3, pp. 9–72 [1959] and Vol. 16, pp. 7–94 [1985]). In the past few years a number of other living monoplacophorans have been identified, including specimens collected in the last century that eluded correct identification until a few years ago. Among the more recent reports are those by V. Urgorri and J.S. Troncoso (*Journal of Molluscan Studies*, Vol. 60, pp. 157–63 [1994]); J. Goud and E. Gittenberger (*Basteria*, Vol. 57, pp. 71–8 [1993]); and A. Waren and S. Hain (*Veliger*, Vol. 35, pp. 165–76 [1992]).

36. An introduction to this interesting group of molluscs may be found in the contribution by A.H. Scheltema, T. Tscherkassky, and A.M. Kuzirian in *Microscopic Anatomy of Invertebrates* (ed. F.W. Harrison and A.J. Kohn), Vol. 5 (Mollusca I), pp. 13–54 (Wiley-Liss, 1994).

37. Although this is the orthodoxy, we need not be surprised if at least one of the two main groups of aplacophorans transpires to be an advanced, derived form that represents a specialized offshoot of more orthodox molluscs.

38. This work is described in *Paleobiology* (Vol. 16, pp. 287–303 [1990]).

39. A useful paper providing a scholarly overview of the annelid–mollusc connection is given by M.T. Ghiselin (*Oxford Surveys in Evolutionary Biology*, Vol. 5, pp. 66–95 [1988]).

40. The detailed descriptions of *Wiwaxia* and the halkieriids are given in two papers in *Philosophical Transactions of the Royal Society of London* B (Vol. 307, pp. 507–86 [1985] and Vol. 347, pp. 305–58 [1995] respectively).

41. In his description of the microstructure of the sclerites of *Wiwaxia* in *Paleobiology* (Vol. 16, pp. 287–303 [1990]) Nick Butterfield was adamant that they were not hollow. Subsequently he has conceded that this claim was in error (personal communication).

42. In our paper on the halkieriids we present evidence for a possibly vestigial shell in a disarticulated specimen of *Wiwaxia*. It needs to be conceded, however, that a scrutiny of other disarticulated specimens from the Burgess Shale did not reveal any obvious shells. In the fauna, from Kaili in Guizhou province, China, that yields disarticulated material of *Wiwaxia* in the form of isolated sclerites (see the paper by Y-L. Zhao and co-workers in *Acta Palaeontologica Sinica*, Vol. 33, pp. 357–66 [1994]), there are also shells ascribed to the genus *Scenella* (see J-R. Mao and Y-L. Zhao in *Acta Palaeontologica Sinica*, Vol. 33, pp. 325–8 [1994]) which are not readily comparable to known *Scenella* and conceivably formed part of the wiwaxiid scleritome. In addition, the recognition of undoubted shells in articulation with Sirius Passet halkieriid has reopened the possibility that various mollusc and brachiopod-like shells were derived from disarticulated scleritomes of related species. In particular, a paper by me and A.J. Chapman in *Journal of Paleontology* (Vol. 71, pp. 6–22 [1997]) explores this proposal with respect to halkieriids from Xinjiang, China.

43. A key paper concerning the position of lophophorates according to data from molecular biology was published by K.M. Halanych and others in *Science* (Vol. 267, pp. 1641–3 [1995]). While not disputing their fundamental conclusion that lophophorates are indeed protostomes, I and a group of molecular biologists based in Glasgow questioned a number of the details (see *Science*, Vol. 272, p. 282 [1996]; see also commentary in *Nature*, Vol. 375, pp. 365–6 [1995]).

44. An interesting paper on the assumed convergence of the lophophore in the protostome lophophorates and deuterostome hemichordates (pterobranchs) is published by K.M. Halanych in *Biological Bulletin* (Vol. 190, pp. 1–5 [1996]).

45. These are well known in Recent brachiopods, but because of their chitinous composition are very seldom preserved in fossil material. Beautiful examples are, however, known from the Burgess Shale brachiopods (see fig. 52 of *Micromitra* in *The fossils of the Burgess Shale* by D.E.G. Briggs and others).

46. A number of papers deal with the similarity of brachiopod setae to annelid chaetae. L. Orrhage in *Zeitschrift für Morphologie, Ökologie und Tiere* (Vol. 74, pp. 253–70 [1973]) provides a useful introduction.

47. Information on the early development of the brachiopod *Neocrania* is given by C. Nielsen in *Acta Zoologica, Stockholm* (Vol. 72, pp. 7–28 [1991]).

48. These include groups known as the serpulids and sabellids.

CHAPTER 8

Other worlds

Rerunning the tape

It is often said that a key event may change radically the whole way in which we see the world and our place in it. The effect does not have to be instantaneous, more often it seeps slowly forward until ultimately our entire world-picture is altered (or distorted). Eventually, if we look backwards we wonder how we could ever have seen things differently. In our time the pictures of our beautiful planet hanging in the utter blackness of space have brought home more forcibly than any other message the fact that we share one Earth. In the past other discoveries must have administered salutary shocks. Contacts with new peoples through maritime expeditions or the recovery of ancient learning were not only traumatic, but sometimes beneficial catalysts.

In its own way, therefore, does not the new understanding of the Burgess Shale inevitably alter the way we view the processes and consequences of evolution? Given that humans are only one of the billions of end points that can be traced from the Cambrian explosion, is it now time to reconsider our own place in the pantheon of evolution? And here too the Burgess Shale has been presented by Steve Gould in his book *Wonderful life* not only as a focus for new ways of looking at evolution, but simultaneously he has elevated this fauna to high importance because of its seemingly profound implications for the processes of historical contingency and hence the place of Man. In this chapter I shall argue that here too several of the claims made by Steve Gould are perhaps exaggerated, and that some of them may be either incorrect or simply uninteresting. Despite the length of the argument in *Wonderful life*, its main strand concerning historical contingency is easy to understand and can be briefly explained. At first sight Steve Gould's line of reasoning hardly seems controversial. Any historical process, be it the history of life over millions of years, or the story of a nation's development over several hundred years, must be riddled with contingent events. Their effect is, he maintains, to render almost any prediction of the future course of history a futile and redundant exercise. It is certainly intriguing, if not entertaining, to imagine how our histories may have unfolded had a saint or megalomaniac died at birth rather than spreading respectively their benign or more often baleful influence.

Writers of science fiction have been especially intrigued by these imaginative possibilities. In his book *The alteration*[1] the English novelist Kingsley

Amis provided an ingenious introduction to a twentieth-century Europe in which there had been no Reformation, and effective power still largely lay with the Catholic Church. In this imaginary theocracy much is familiar: Mozart was, for example, born in 1756. (Although Kingsley Amis does not mention it we could, perhaps, speculate that in his imaginary world Mozart never wrote the *Magic Flute*, for this opera draws heavily on the precepts of Freemasonry, which an all-powerful Church would surely have suppressed.) In the book, Mozart did not die in 1791, but lived to a ripe old age in which he composed a whole series of masterpieces, including a second *Requiem*. Half the fun of *The alteration* is the continued skewing of the familiar. For example, instead of capturing an aspect of eternity in his transcendental watercolours, the genius of the English painter J.M.W. Turner is expressed in the superb ceiling paintings of a mighty English cathedral. The Pope, John XXIV is, of course, Supreme Pontiff in Rome, but he is a blunt-speaking Englishman from Yorkshire. His advisers include Francis Crick, mentioned only in passing as a malign figure in charge of drastic methods of birth control.

It is all very entertaining, and why shouldn't an evolutionary biologist write a book in the same spirit as Kingsley Amis's *The alteration*, but about an alternative history of life? Not surprisingly, at least one such book has been written. This is Dougal Dixon's playful exploration of evolutionary possibilities in a book published in 1981 and called *After Man: a zoology of the future*.[2] He asks his readers to envisage a world that is emerging from an earlier ecological catastrophe. In the book he supposes that of all the mammals only a handful or types, mostly rats and rabbits, survived to re-populate the globe. *After Man* is an exercise rich in imagination in its depiction of the riot of species that quickly radiate to refill the vacant ecological niches left after a time of devastation. All the animals, of course, are hypothetical. Certainly they look very strange, sometimes almost alien. When we look more closely at their peculiarities, however, they turn out to be little more than skin-deep. In this imaginary bestiary the basic types of mammal, those that trot across the grasslands, burrow in the soil, fly through the air, or swim in the oceans, all re-emerge.

The book *After Man* was published some years before *Wonderful life*. It is notable that its fundamental message, that even with an effectively clean slate the re-emergence of new forms of life has a basic predictability, is not addressed in *Wonderful life*. Indeed, this book appears not even to be mentioned in *Wonderful Life*. In contrast, Steve Gould argues passionately that were we 'to rerun the tape of life' from the time of the Cambrian explosion, we would end up with an utterly different world. Among its features would be an almost certain absence of humans or anything remotely like us. At first sight, this seems to be a beguiling scenario. In fact, as stated, I consider the metaphor of rerunning the tape of life to be a rather trivial exercise, worthy of only passing mention. Much more importantly, its premise is based on a fundamental misapprehension of both the nature of the Burgess Shale fauna and the processes of organic evolution. Furthermore, the apparently radical claims in *Wonderful life* may actually conceal some much more interesting problems.

First, however, let me explain why I think that the metaphor of rerunning the tape of life is rather trivial. Viewed biologically, one hundred years ago my existence would have been inconceivable other than perhaps as the fond but hypothetical musings of my grandparents. If my parents had not met, the world would be full of humans, but I would not be writing these words. But it is worse than that. If my parents had not made love on a particular day in early February 1950, again I would not be here. Their child, conceived on another occasion, would certainly be similar to me, but distinguishable in all the ways that brothers and sisters do differ. And so this argument also stretches back through my and your long chain of ancestors, whose meetings and conception of children were matters of chance and accident. But does this really matter? If Charles Walcott had not discovered the Burgess Shale, sooner or later another geologist would have done so. If I had not had the good fortune to work on the Burgess Shale, sooner or later another palaeontologist would have done so.

Is the historical path leading to a species any different from that leading to an individual human? It really is rather pointless to say from the perspective of hypothetical observers in the Cambrian that the rise of humans would be effectively unpredictable. So many species had to evolve and give rise to other species, in a chain stretching through half a billion years. So tortuous is the path from an animal like *Pikaia* to a human via all sorts of fish, amphibians, reptiles, and mammals that there does indeed seem to be no predictability in the outcome. Indeed, when in the history of evolution could we say that the emergence of modern humans was inevitable? Probably not as recently as four million years ago, when our australopithecine ancestors could be regarded as just another type of ape; not necessarily with the evolution of the genus *Homo*. Research is now beginning to indicate that there was a whole plethora of early species, all now extinct together with their primitive stone tool kits. Rather surprisingly, perhaps not even a mere 150,000 years ago, when anatomically modern *Homo sapiens* evolved. This is because in the first half of our history the behaviour of our species seems to be strangely conservative and unimaginative. The real breakthrough in terms of Palaeolithic art and hunting technology occurred only about 50,000 years ago.[3] It seems to have been astonishingly rapid and as yet lacks a convincing explanation.

Surely this whole argument, focusing on the implausibility of humans as an evolutionary end product, misses the point. It is based on a basic confusion concerning the destiny of a given lineage, be it of a human family or a phylum, versus the likelihood that a particular biological property or feature will sooner or later manifest itself as part of the evolutionary process. The point is that while the former, say the evolution of the whales, is from the perspective of the Cambrian explosion no more likely than hundreds of other end points, the evolution of some sort of fast, ocean-going animal that sieves sea water for food is probably very likely and perhaps almost inevitable. Although there may be a billion potential pathways for evolution to follow from the Cambrian explosion, in fact the real range of possibilities and hence the expected end results appear to be much more restricted. If this is a correct diagnosis, then evolution cannot be regarded as a series of untrammelled and unlimited experiments. On the contrary, I believe it is necessary to argue that within certain limits the outcome of evolutionary processes might be rather predictable. Let us think back to the hypothetical imaginings of the book *After Man*. To our human eyes many of the newly evolved mammals described in this book look very strange: dream-like, or perhaps even the stuff of nightmares. A closer and more critical look reveals that the basic types of ecology are little different from those of the familiar mammals of the present-day world. One might argue that the animals depicted in *After Man* have the basic similarities in terms of their ecology to those of living mammals simply because Dougal Dixon has failed to imagine alternatives. Is it possible to accept such an assertion? To refute this question it is necessary to introduce the evolutionary phenomenon known as convergence. This is the phenomenon that animals (as well as plants and other organisms) often come to resemble each other despite having evolved from very different ancestors. Nearly all biologists agree that convergence is a ubiquitous feature of life.[4] Convergence demonstrates that the possible types of organisms are not only limited, but may in fact be severely constrained. The underlying reason for convergence seems to be that all organisms are under constant scrutiny of natural selection and are also subject to the constraints of the physical and chemical factors that severely limit the action of all inhabitants of the biosphere. Put simply, convergence shows that in a real world not all things are possible.

Some examples may help to clarify this discussion. The book *After Man* invites us to enter a new world, filled with a dazzling diversity of mammals that arose for the most part from the seemingly unremarkable rabbits and rats. In fact, similar experiments in the natural world have already taken place. Here is one example. For much of the Tertiary, that is the period of geological time that began about 65 Ma ago, South America was isolated, surrounded by oceans. It was a sort of super-island. Before South America became cut off there were already indigenous mammals there. Once isolated, South America turned into a sort of natural laboratory as the mammals underwent a rapid evolutionary radiation.[5] An enormous variety of forms appeared at one time or another such as giant sloths, weighing almost four tonnes, and armadillos as big as a military tank (the glyptodonts). Some of the parallels with mammals that evolved elsewhere are really very striking. Perhaps the most famous example is a sabre-tooth 'cat'. It is very similar to the sabre-tooth tigers that were found across much of the Northern Hemisphere until recently. But despite the fact that both animals have massive canines for stabbing their prey (Fig. 87), their origins are very

Fig. 87. A classic example of convergent evolution. The lower animal is the famous sabre-tooth tiger. The upper animal looks very similar, but it is actually a marsupial. Its giant stabbing canines evolved completely independently, but for a very similar purpose. [Reproduced with permission of Marlene Hill Werner (The Field Museum, Chicago), from fig. 10 of L.G. Marshall (1981). The great American interchange—an invasion induced crisis for South American mammals. *In* M.H. Nitecki (ed.) *Biotic crises in ecological and evolutionary time*, pp. 133–229. Academic Press, Orlando, Florida.]

different. The sabre-tooth tiger of the Northern Hemisphere is a close relative of the living tiger and panther. In contrast the South American example is nothing to do with the big cats. It is a marsupial, related to the living kangaroos and opossums.

In South America most of the indigenous mammals were driven to extinction several million years ago, when a connection was established with North America.[6] According to the traditional view, a flood of invading species at this time apparently sealed the fate of the South American animals; much of the living fauna (llamas, jaguars, etc.) is actually composed of North American immigrants. But not all the indigenous South American mammals were doomed. Armadillos, for example, not only remain widespread in South America but have migrated into the southern states of the United States. The marsupial opossums have done even better: they have reached as far north as Toronto.

On another super-island, Australia, the indigenous mammal fauna still flourishes, although it has taken some hard knocks from the recent invasions by Man and some co-placentals such as dogs, rats, and cats. Here, too, there are some striking instances of evolutionary convergence. In the Northern Hemisphere a common mammal is the mole. In Australia there is a marsupial equivalent, also equipped with powerful forelimbs for digging and with only weak eyes in keeping with its subterranean existence in the sands of the outback. Marsupials and placentals are both mammals and so share a common ancestor, which must have lived about 100 Ma ago when the two lineages split. But the placental mole and marsupial mole (and the similar example of the sabre-tooths) evolved independently towards surprisingly similar end points.

It is important to stress that these parallels in animal 'design' are very seldom precise. In addition, because a certain type evolves in one place, this is not a guarantee that a convergent equivalent will emerge somewhere else. In South America, for example, giant birds (the phorusrhacids) grew almost 3 m high. They are now extinct, but appear to have been efficient hunters and killers. Giant birds are certainly known from other areas of the world (think of the ostrich of Africa or the extinct moa of New Zealand), but they are not very similar to these South American birds. In Australia marsupial carnivores evolved, but perhaps because or the climatic conditions they never seem to have reached the size or ferocity of the placental mammals as exemplified in Africa.

Nevertheless, the point I wish to stress is that again and again we have evidence of biological form stumbling on the same solution to a problem. Consider animals that swim in water. It turns out that there are only a few fundamental methods of propulsion. It hardly matters if we choose to illustrate the method of swimming by reference to water beetles, pelagic snails, squid, fish, newts, ichthyosaurs, snakes, lizards, turtles, dugongs, or whales;

we shall find that the style in which the given animal moves through the water will fall into one of only a few basic categories.

Does it follow then that contingent processes are an irrelevance in the way we see the world? I have argued that, so far as the history of life is concerned, they are. But I also believe that there is a special and unique exception, relevant only to human history. I am not, of course, referring to novels or other works of fiction that portray an alternative history. However strange the depiction of an imaginary world may be, however ingenious the imagination of the author, somewhere or other a parallel will be found in the real world. The long history of mankind is studded with convergences, perhaps most notably in social systems and the use of artefacts and technology. But for human history, set in the arrow of time, there appears to be one intolerable stumbling-block. This is the catastrophic failure in human values and decency. The list is almost endless: the sacking of Constantinople in 1204, the destruction of Baghdad in 1258, the expulsion of the Jews from Spain in 1492, and the Nazi Holocaust are only some among the infamous epochs in the litany of disaster. If there were a clear prospect that such evils were part of a barbarian past, then at least we might find a small crumb of comfort. No such prospect exists: no scientific analysis can even remotely answer or account for past and present horrors of human behaviour. It is my opinion that human history can make no sense unless evil doings are recognized for what they are, and that they are bearable only if somehow they may be redeemed.

How weird is the Burgess Shale?

I hope that by now I have persuaded you that whatever importance is attached to the Burgess Shale, it is not in the operation of either historical contingency or in the fable of re-running the film of the history of life. It really is not very important if the many details of an alternative history are different, because in broad outlook the study of evolutionary convergence demonstrates that the world, perhaps even any world, would have to look broadly similar.

So let us finally return to all those apparently strange and bizarre animals in the Burgess Shale. It is the belief of Steve Gould that these fossils could well undermine our conventional view of evolution. This is not the first time that Steve Gould has suggested that the existing edifice of evolutionary theory is ripe for demolition, yet (as I noted in the first chapter) when the dust, heat, and smoke have eventually died down once again little seems to have changed. Why then should we regard his ideas concerning the Burgess Shale with scepticism? His first argument, that the problematic fossils have such distinctive body plans that they must represent extinct phyla, is misleading.

As we saw above, supposed problematic fossils, such as the wiwaxiids and anomalocarids, actually appear to be crucial to our understanding of how known phyla evolved in their earliest stages. The second point, also explained above, is that there seems to be little need to invoke radical new mechanisms of evolution to explain the origin of these supposedly bizarre groups. The reasons for such scepticism come first from our steadily improving knowledge of the fossil record. Species supposedly isolated on account of their apparently weird anatomy are now beginning to be placed in a sensible framework of evolution. The second point concerns the basic similarities between animals in the genetic mechanisms of development. As we saw in Chapter 6, if a fly and a mouse have body plans that are laid down by the same fundamental instructions, then it is perhaps less likely that the self-evident differences in anatomy stem from really radical differences in genetic architecture.

The concept of disparity

I mentioned earlier in this chapter that although the discussion of contingency with respect to the Burgess Shale is really rather trivial, there do remain two more significant questions. The first is well within the realm of evolutionary biology, and asks what are the limits to biological design? This chapter will conclude with a brief discussion of this point. The second point is really metaphysical, and poses the following question. Even as a product of evolutionary processes can we as humans make any claim to uniqueness? It will be on this subject that this book will be concluded.

It is probably true to say that the majority of zoologists, if asked to describe in a couple of sentences the hundreds of millions of years of animal evolution, might reply more or less as follows: 'Now, don't quote me, but I suppose the first animals were pretty simple, and now look at the planet. It is stuffed with an amazing variety of things: parrots, sea anemones and so on'. If you then asked this zoologist to sketch such a concept, he or she might draw a sort of cone, standing on its pointed end (Fig. 88(a)). If the discussion continued, it would probably be agreed that it might be sensible to distinguish between sheer numbers of species—what might be called diversity—and range of different types or designs forms—what is now generally referred to as disparity. This important distinction, between diversity and disparity, has been articulated by a number of palaeontologists, including Bruce Runnegar.[7]

How then can we tackle the problem of measuring disparity, the nearest approximation we have to comparisons of biological 'design'? For closely related groups, comparisons can be fairly straightforward. Consider ourselves and the chimpanzees. Just by comparing the anatomies of the skeletons, it will be clear that the differences are mostly ones of proportions: shape of skull, size of teeth, lengths of limbs, and so forth. Suppose that we wish to

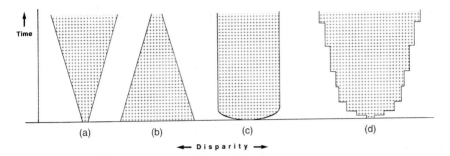

Fig. 88. Various interpretations of the history of life and its disparity. (a) The traditional view whereby disparity steadily increases through geological time. (b) The view presented by S.J. Gould, whereby maximum disparity is in the Cambrian. (c) The view that disparity increased very rapidly in the Cambrian and thereafter stayed much the same. (d) The view presented here. Disparity increased rapidly in the Cambrian, and thereafter has generally increased, albeit at varying rates.

expand our range so as to include cows and bats. As these are also placental mammals with the same basic type of skeleton, comparisons are still possible, but the difficulties in terms of meaningful results are becoming formidable. If we wanted to include representatives of other phyla, say a shrimp and a slug, then any objective comparison of anatomy would be practically impossible.

Overall the evidence of diversity increasing through geological time is rather convincing.[8] Even though the fossil record is highly inadequate in terms of species, the higher unit (a family) within the taxonomic hierarchy appears to provide a reasonable proxy (Fig. 89). What about disparity? Here

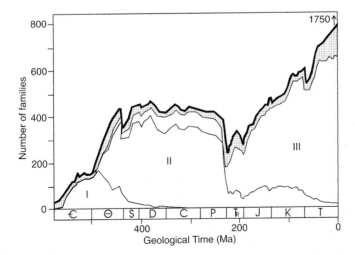

Fig. 89. The increase in diversity of marine animal life since the Cambrian. Symbols denote geological periods: Cambrian, Ordovician, Silurian, Devonian, Carboniferous, Permian, Triassic, Jurassic, Cretaceous (K), Tertiary. [Based with permission on fig. 5 of J.J. Sepkoski (1981). A factor-analytic description of the Phanerozoic marine fossil record. *Paleobiology*, Vol. 7, pp. 36–53. Paleontological Society]

we enter much more controversial waters, with the argument largely stirred up by conflicting interpretations of the Burgess Shale fauna. By this time in the Cambrian we have evidence for animals as different as ctenophores, arthropods, and chordates, not to mention other animals that apparently look so extraordinary that according to some people there is only one option: place them in hitherto unrecognized phyla. If there are about 35 phyla recognized today, perhaps in the Cambrian there are several hundred! It is in this general context that Steve Gould makes what at first seems to be a reasonable claim as part of his extended argument on the Burgess Shale. He does this by proposing that there is an enormous asymmetry in the extent of disparity through geological time. Employing the metaphor of a cone of life, he would invert it (Fig. 88(b)) so that maximum disparity was attained in the first stages of animal evolution and thereafter was progressively reduced during geological time. This would approximate to a steady reduction in the number and variety of body plans, so that in some sense today's fauna is an impoverished remnant of the former glories of the Cambrian. It comes as no surprise that this view goes hand in hand with the notion, already discussed, of historical contingency. After all, if the cone of disparity is constantly narrowing, if more and more animal designs are being weeded out by whatever mechanism of extinction, then the likelihood of anything remotely similar happening again in our actual history is that much lower. A beguiling scenario, but is it correct? First, it is worth realizing that the concept of 'inverted' or 'upright' cones of disparity are not the only possibilities. For example, it is conceivable that the disparity in modern oceans is much the same as it was in the Cambrian (Fig. 88(c)). Alternatively, it could be the case that disparity has indeed increased through geological time, but in a step-like manner rather than in a regular fashion (Fig. 88(d)). In this case one could envisage occasional innovations in evolution, the consequences of which ripple through the biosphere and so drive times of rapid change.

It is also important to stress that different disparities may need to be disentangled. Obviously the disparity of trilobites has been zero since 250 Ma ago, because that was when they became extinct. Indeed, as we shall see, the changes in the disparity of trilobites through time are now rather well mapped. It would similarly not be unreasonable for other groups to show either an overall decline in disparity, or alternatively an increase. An example of the former might be represented by the priapulid worms. The evidence available at the moment suggests that in the Cambrian period they showed a wide range of designs. Unfortunately their fossil record is rather poor after the Cambrian, but it looks as if this group might today be rather impoverished in comparison with their heyday in Burgess Shale times. On the other hand, the bivalve molluscs (known today from animals such as oysters and scallops) appear to have a far greater disparity today than in the Cambrian. In fact, much of the discussion concerning disparity has revolved around the arthropods, which is not at all surprising given that they are both the most

abundant group in the Burgess Shale-type faunas and also the most frequently found skeletal fossils in the Cambrian. Arthropods are thus taken as
a proxy for the entire marine realm, and this seems reasonable.

To consider the problem of disparity let us first look at the trilobites.
We encountered trilobites in the Burgess Shale in the form of *Olenoides* (Fig.
37), where they are of special interest because of the superbly preserved
appendages. These are rather delicate, at least in terms of surviving decay,
and so are very seldom fossilized. Thus, when studying trilobites palaeontologists need to rely almost exclusively on the exoskeletons. These are composed
of calcium carbonate and are very common as fossils for two reasons. First,
the exoskeleton is very resistant to decay and so readily fossilizes. Second,
during the life of the trilobite the animal faces a simple dilemma. It continues
to grow, but it lives in a rigid exoskeleton than can neither stretch nor
expand. The solution, which is one of the characteristics of arthropods, is to
break out of the exoskeleton in a process known as moulting.[9] Having done
so it then resecretes a new and larger skeleton. Thus during its life a trilobite
will produce a number of separate moults, each a potential fossil.

To a first approximation the exoskeleton of the trilobite consists of three
units: large shields at either end of the body covering respectively the head
and tail, and an intervening series of narrow segments. Upon moulting or
death this exoskeleton usually breaks up, and so the head-shield (called the
cephalon) is often found isolated. There is one further complication. In many
trilobites the process of moulting is assisted by two small plates on the front
of head becoming detached (Fig. 90). In some unusual fossil trilobites the

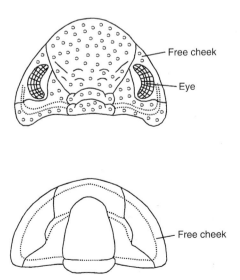

Fig. 90. Typical trilobite heads. The two small plates (the free cheeks) often detach
from the remainder of the head.

exoskeleton is preserved in the post-moult configuration, although the trilobite itself has crawled away. In such examples these two small plates may be seen to have been flipped off on either side of the head.

Mike Foote, now in the University of Chicago, decided to study the fossilized heads of a large number of trilobite species.[10] In his first analysis he looked at about 560, stored in various museums. It is quite difficult to find complete specimens of these heads, but if at least half the head was visible, that was enough: the other half is, of course, a mirror image because trilobites are bilaterally symmetrical. In ideal circumstances, at least for a palaeontologist, one would study the changes in the shapes of the trilobite heads at fairly regular time intervals, say every 1 Ma. This is not yet feasible, but it appears to be legitimate to take a sort of average for longer periods of time. Thus, Mike Foote chose six slices of geological time: three for the Cambrian period (Lower, Middle, and Upper) and an equivalent three for the succeeding period, the Ordovician. In sum, this represents an enormous period of time, about 100 Ma. In this geological interval trilobites are both abundant and occur in a wide variety of forms. After the Ordovician comes the Silurian, and in this period trilobites are still quite frequent. Thereafter they go into a long-term decline.[11] Thus, although trilobites survived until the end of the Palaeozoic they are generally rather rare for the last 165 Ma of their history.

Although built to a basically identical plan, in detail the heads of trilobites come in a considerable variety of shapes (Fig. 91). We therefore need a method of some sort that will enable us to make an objective comparison of all these different shapes. In addition, we might well expect that the extent to which morphospace is occupied will vary in some way over periods of many millions of years. The actual methods used are sophisticated, and to apply them in detail requires a knowledge of statistics and computing. The aim is to describe mathematically the total range of morphological types. Trilobite heads are quite complicated in shape, and to analyse the morphospace they occupy it is usually necessary to rely on many measurements. In doing so one is defining a volume of a kind, which because it is based on many measurements is not restricted to the familiar three dimensions: when assessing mathematically large numbers of measurements we need to use equations that define a multidimensional space. In this way one can define a region of morphospace that mathematically describes the particular morphologies—in this case a wide range of trilobite heads. Because the human mind cannot readily envisage such concepts, it is usual to employ special computer programmes that effectively circumvent this problem. Using statistical techniques such as Principal Components Analysis (PCA), these multidimensional data can be 'collapsed' back into two or three dimensions and so plotted as graphs in a comprehensible fashion.

Now let us return to the problem of defining mathematically the shape of a trilobite head.[12] In principle, the outline shape is plotted on a digital table so

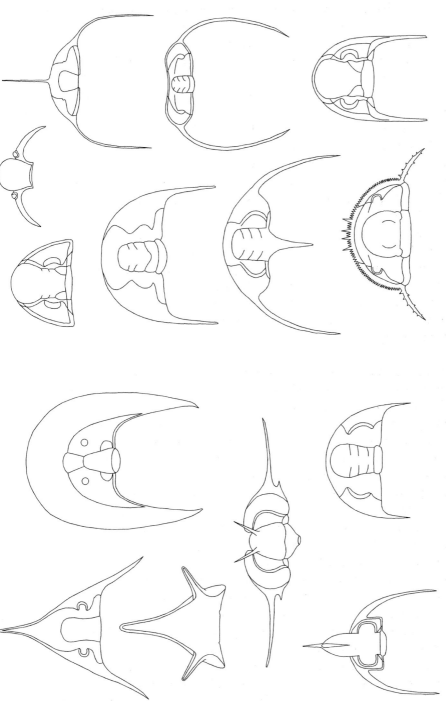

Fig. 91.   A selection of trilobite heads. Although they share a basic design, in detail they are widely different.

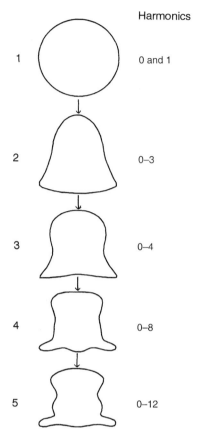

Harmonics

1   0 and 1

2   0–3

3   0–4

4   0–8

5   0–12

Fig. 92. Defining a trilobite head. [Based on fig. 4 of M. Foote (1989). Perimeter-based Fourier analysis: A new morphometric method applied to the trilobite cranidium. *Journal of Paleontology*, Vol. 63, pp. 880–5. Paleontological Society.]

that the information can be entered directly into a computer. What the computer programme effectively does is take a series of ellipses, the harmonics of which in combination will reproduce almost exactly the outline shape of the original head. Obviously if one tried to describe the shape of the head on the basis of a single ellipse it would be very poor approximation. However, if one adds a succession of ellipses it does not take long to provide a very passable imitation of the outline of the head (Fig. 92). Mathematically this series of ellipses can then be described by a series of harmonic equations, which in effect are based on the angular relationships (expressed as sines and cosines) defined by the sweeping out of the edge of the trilobite head by the line that connects the central point to the perimeter. Each species of trilobite, of course, has a different shape of head. If the various heads are similar, then obviously the amount of morphospace occupied will be rather small. Conversely if there is a wide range of shapes, in other words if there is major disparity, then a much larger volume of morphospace will be occupied.

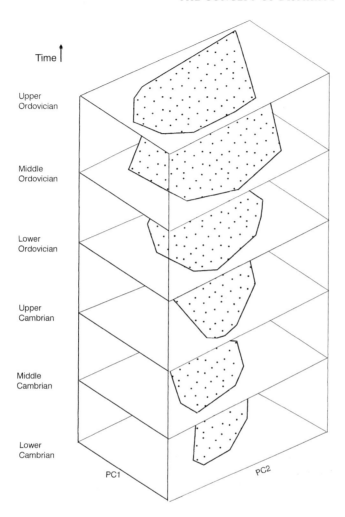

Fig. 93. The occupation of morphospace by trilobite heads through six intervals of geological time, between the Lower Cambrian (*c.* 530 Ma ago) and the Upper Ordovician (*c.* 440 Ma ago). [Data based on fig. 2 of M. Foote (1991). Morphologic patterns of diversification: examples from trilobites. *Palaeontology*, Vol. 34, pp. 461–85. Paleontological Association.]

The results of the analysis by Mike Foote are most interesting. Looking at each of the time-slices he found that the occupation of morphospace steadily increased from the Lower Cambrian to the Middle Ordovician (Fig. 93). Only in the Upper Ordovician was this trend reversed so that the occupation of morphospace began to diminish. What we do know is that at about this time the trilobites entered a mass extinction, and then a long-term decline that ended with their final extirpation at the end of the Palaeozoic. Mike Foote subsequently extended this analysis of trilobite disparity,[13] considering both the entire history and the relative disparity of particular groups such as the phacopid trilobites, a large and successful assemblage notable for their prominent compound eyes. In this extension of his earlier work he confirmed that after the Ordovician period trilobite disparity did indeed decline. Nevertheless, at first this was relatively slow, and became precipitous only after the end of the Devonian, that is about 80 Ma after the peak in disparity.

There are further questions we can ask about the occupation of morphospace. In particular, is it filled in a fairly uniform pattern or are there a series of discrete clusters separated by zones of unoccupied morphospace? When Mike Foote initially analysed this problem[14] he found that the answer varied according to the point in geological time that was chosen. In fact one could perhaps draw a rather crude analogy with the evolution of the Universe. There is an evolutionary Big Bang and the trilobites appear (as part of the Cambrian explosion). At first the area of morphospace occupied is rather small, and moreover the distribution of the trilobite species is rather homogenous, without obvious clustering. This would be very roughly analogous to the 'soup' of elementary particles shortly after the Big Bang. Later in geolo-

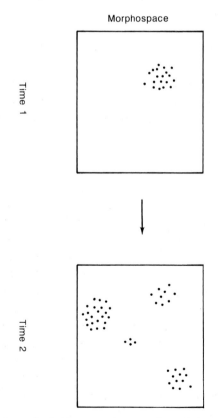

Fig. 94. The occupation of trilobite morphospace by trilobites at an early stage of their evolution and at a subsequent stage. [Concept based on fig. 1 of M. Foote (1990). Nearest-neighbor analysis of trilobite morphospace. *Systematic Zoology*, Vol. 39, pp. 371–82. Society of Systematic Biologists]

gical time, in the Ordovician, the net extent of morphospace has grown dramatically. When we look at the distribution of the trilobites, however, there is now a series of discrete patches separated by voids where there are very few, if any, trilobites (Fig. 94). The analogy with the history of the Universe would now be the stage when the galaxies form and are separated by more or less empty space. Beyond this the analogy begins to break down. Admittedly trilobite morphospace then begins to contract, as one day the Universe may contract, but with the contraction of morphospace there is no reamalgamation into some sort of primal state. Nevertheless, the details of this decline in disparity are interesting. In particular, at first the weeding out of morphospace appears to have been rather unselective. Hence, during the initial decline, that is during the Silurian and Devonian, the steady loss of species does not make much of an impact on the total volume of occupied morphospace. This process of attrition, however, could not continue indefinitely and the later trilobites have rather generalized morphologies that display nothing like the morphological exuberance of the earlier representatives.

These analyses by Mike Foote are very elegant, although in one sense they only reiterate facts long known to trilobite palaeontologists. It had been appreciated for many years that Cambrian trilobites were by and large a rather homogenous crowd and it was not easy to subdivide them into a series of obviously natural groupings. In contrast, Ordovician trilobites fall into a whole series of distinct groups with little in the way of intermediates connecting them. Thereafter, as just noted, the trilobites enter their long decline, both in terms of disparity and taxonomic richness. By the Carboniferous they are a pretty dull lot.

The statistical analysis by Mike Foote thereby placed in a rigorous context what had been intuitively understood for a long time. Most importantly, it shows that so far as trilobites are concerned, or least their heads, there is an increase in disparity through geological time. Until the end of the Ordovician it is an 'upright cone' (Fig. 88(a)). The subsequent contraction in morphospace is most probably due to competition with other groups in the marine realm.[15] The emergence of discrete groupings (Fig. 94) also poses some interesting questions. If the history of trilobites could be rerun from the beginning of the Cambrian there might be similar, but not identical, occupation of morphospace. Perhaps the discrete zones of occupation would shift, so that new types of trilobite would emerge from the sea of contingent possibilities. But perhaps not. It could turn out to be the case that the areas occupied are by no means accidental. It is conceivable that some 'designs' of trilobite head are better adapted, perhaps by avoiding structural weakness or determining the arrangement of sensory organs. In this case the zones occupied might correspond to points of relative stability.

Trilobites are only one group of arthropods, and as we have seen in the Burgess Shale and similar faunas they are not especially important. What about arthropods as a whole? How does their disparity change through geological time? The difficulties of assessing this problem are formidable. The most obvious obstacle is that whereas comparing the various shapes of trilobite heads is relatively straightforward, such techniques simply cannot be applied to the enormous range of arthropod body plans. Though they evidently come from a common ancestor, even by the Lower Cambrian the arthropods are simply too different. Appendages, presence of a carapace, number and type of segments, and so forth simply show too many differences. In other words there is no single scheme of comparison, at least in terms of the morphometric analysis employed by Mike Foote in his study of trilobite heads, that can be applied. This does not mean that objective comparisons are impossible. Any arthropod, including those of Burgess Shale type, possess a large number of morphological features. They are, after all, very complex animals. The usual procedure is to make a long list of these features. This is a fairly objective exercise. Each item of morphology is termed a character state. It may be recorded as something which is either present or is absent. Alternatively, the feature may be invariably present but occur in a number of different states. For example, eyes may be present or absent. On the other hand, all arthropods have appendages. At a fairly basic level these might occur as lobopods (as in the Burgess Shale *Aysheaia*), as a leaf-like appendage (these are particularly characteristic of a group of crustaceans known as the branchiopods, such as the living *Triops*), or as the classic jointed appendage (such as we encountered in the Burgess Shale *Marrella*, Fig. 36).

How then is it possible to use the enormous amount of information on arthropod character states to decide whether the amount of morphospace occupied by Cambrian arthropods was similar to or vastly different from that occupied by living arthropods? An interesting analysis was undertaken by Matthew Wills.[16] At the time he carried out this work, he was a research student in the University of Bristol, collaborating with Derek Briggs. The details are rather technical, but the outlines of the argument and the conclusions that Wills drew are quite straightforward. In general it is believed that the longer the list of character states is, the better will be the analysis.[17] Matthew Wills used 59 basic character states. Many were simple alternatives (present versus absent), but some had up to six variants. He examined their distribution in equal numbers of Cambrian and living arthropods, 24 of each. In some arthropods, of course, a particular character state remains unknown. Most frequently this is a fossil species, where preservation is incomplete or too poor to resolve definitely the structure in question. Nevertheless, such gaps in the database are quite sparse. With 59 character states and 48 species of arthropod the resultant matrix has 2832 boxes to be filled. Of these only 174 contained question marks indicating an unknown character state; that is, a mere 6 per cent.

The technique employed involves running the data through a fairly sophisticated computer programme for statistics. The basic aim, however, is quite straightforward. Let us suppose that two taxa share a large number of character states. It is obvious that between them they can occupy only a limited amount of morphospace. The converse will apply, of course, if two taxa have almost no character states in common. Each taxon would then occupy a point in morphospace far removed from the other. Comparing any two taxa is quite straightforward, but when it comes to analysing all the Cambrian arthropods (or alternatively all the living examples), the amount of data is impossibly large for a human brain to manipulate.

The computer programmes employed are similar to those that Mike Foote used in his study of trilobite heads and their occupation of morphospace. The net result of Matthew Wills's analysis is shown in Fig. 95. In Mike Foote's the data were 'collapsed' into two dimensions, but in Matthew Wills's analysis three dimensions are used. Each taxon occupies a discrete point because it

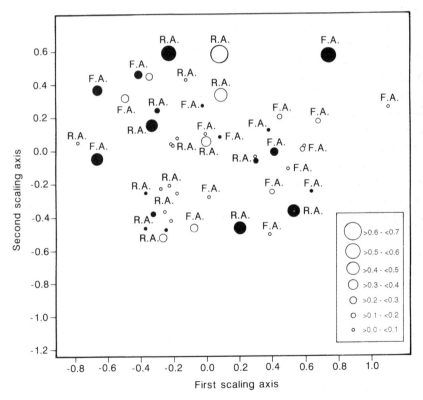

Fig. 95. A comparison of occupation of morphospace by Cambrian and living arthropods. Score on the third scaling axis: solid circles, positive values; open circles, negative values; F.A., fossil arthropods; R.A., Recent arthropods. [Diagram based on fig. 8 of M.A. Wills, D.E.G. Briggs, and R.A. Fortey (1994). Disparity as an evolutionary index: a comparison of Cambrian and Recent arthropods. *Paleobiology*, Vol. 20, pp. 93–130. Paleontological Society.]

differs in at least one character, and usually many more, from all the other taxa. Its position is plotted relative to the first two dimensions on the x and y axes of the graph. The position of each taxon in the third dimension (the z axis) is indicated by the diameter and type (solid or open) of the circle, which shows whether the particular taxon lies above or below the x–y plane.

So much for the technicalities. What is the net result of this analysis in terms of occupation of morphospace? The diagram indicates that the degree of occupation by Cambrian arthropods is much the same as that of living arthropods. In other words their disparities are more or less equal. Steve Gould's claim that the disparity of Cambrian arthropods was substantially greater appears to have been neatly refuted. But the story has probably not yet ended.[18] It has been pointed out that although the number of taxa used in the comparison is equal (24 each), those from the Cambrian may be a much less representative sample than the living examples. On the other hand, the method of scoring morphological characters is relatively crude, and there may be other methods of assessing the morphological richness of arthropods that could lead to a better definition of morphospace occupation. Graham Budd, in the University of Uppsala, is investigating this area and he tells me that some evidence suggests that the disparity of living arthropods is substantially greater than it was in the Cambrian.

The debates about the role of contingency in evolution, the controlling hand of convergence, and the measurement of organic disparity will continue for many years.[19] All are areas of active research, but it is now time to consider, albeit very briefly, the place of Man. We may be a product of evolution, but assuredly we have the ability to transcend our origins. Whether we like it or not, we are a unique species. We have unique responsibilities, even though too often our actions are disastrous. But supposing that we are not only unique to this planet, but are genuinely alone?

Notes on Chapter 8

1. *The alteration* was re-published by Penguin Books in 1988. Two other books that explore somewhat similar themes are *Pavane* by Keith Roberts (Victor Gollancz, London, 1984, re-published by VGSF in 1995) and Philip K. Dick's *The man in the high castle* (Penguin Books, Harmondsworth, 1965). The former has echoes of *The alteration*, dealing with a theocratic state whose grip on power is an attempt to curb the worst excesses of materialism and give time for Man to develop a moral maturity commensurate with his technological prowess. Given our status as spiritual dwarfs and the disasters of materialism, *Pavane* deserves to be better known. Dick's book is much bleaker, and has an uneasy ring of truth.
2. This book, now out of print, was published by Granada, London.
3. Not surprisingly this is a controversial area, and in part depends on the dating of new finds in areas such as the Congo (formerly Zaire) and Australia, which if accepted would suggest cultural innovation tens of thousands of years before the main Palaeolithic revolution. Even more astonishing in this regard is the discov-

ery of hunting spears and associated evidence for butchery in German deposits that are dated at an extraordinary 400,000 years BP (see the paper by H. Thieme in *Nature*, Vol. 385, pp. 807–10 [1997]). Confirmation of this date is urgently required. Some slight evidence is also emerging for powers of symbolic representation in the Neandertals, although some of this may represent imitation of the cultural products of Upper Palaeolithic *Homo sapiens*. The generally accepted evidence for deliberate burial of the dead by Neandertals, however, would perhaps indicate a shared belief in the afterlife, or according to materialists a shared delusion.

4. For an excellent review of convergence in the invertebrates see the paper by J. Moore and P. Willmer in *Biological Reviews of the Cambridge Philosophical Society* (Vol. 72, pp. 1–60 [1997]). Other thoughtful essays on the ubiquity of convergence and the constraints of evolution are by D.B. Wake (*American Naturalist*, Vol. 138, pp. 543–67 [1991]) and B.K. Hall (*Evolutionary Biology*, Vol. 29, pp. 215–61 [1996]). Despite the wide-ranging nature of these articles I am not aware of a single synthesis, and perhaps one is not really necessary: after all rampant convergence is not in dispute and piling up example after example might exhaust rather than instruct. Two additions to this catalogue are, however, particularly noteworthy. The first is by G.C. Williams (*Proceedings of the California Academy of Sciences*, Vol. 49, pp. 423–37 [1997]), who describes a new type of soft coral, strongly convergent in general form on a number of other deep-sea animals. The second is a mechanical analysis of a group of burrowing amphibians known as caecilians by O'Reilly and others (*Nature*, Vol. 386, pp. 269–72 [1997]). This reveals an astonishing degree of convergence to the hydrostatic architecture seen in various burrowing invertebrate worms, despite employing a vertebrate body plan. The constraints of convergence and the strong adaptive advantages of particular configurations are also a recurrent theme in mainstream evolutionary thinking. A good example may be found in the paper by J.P. Hunter and J. Jernvall in *Proceedings of the National Academy of Sciences, USA* (Vol. 92, pp. 10718–22 [1995]), where they discuss a tooth cusp, known as the hypocone, as a convergent and functionally significant character. The constraints on life and the ubiquity of convergence may reopen in an interesting fashion the argument for Design. It is sometimes stated that particularly complex structures, including biochemical pathways, are so intricate in their construction that it is inconceivable that they could have evolved more than once. In certain cases, at least, such as that concerning the development of C4 photosynthesis, independent innovation of the same pathway does appear to be the norm. In all these discussions there is, of course, the risk of circular reasoning inasmuch as evolutionary relationships are established on shared similarities, whereas other features are established as only similar because they are convergent.

5. For a summary of the evolution of mammalian faunas in South America and the Great American Interchange resulting from the formation of the Panamanian Isthmus connecting the Americas see L.G. Marshall in *Biotic crises in ecological and evolutionary time* (ed. M.H. Nitecki), pp. 133–229 (Academic Press, New York, 1981). Other papers relevant to this topic include a review by L.G. Marshall in *American Scientist* (Vol. 96, pp. 380–8 [1988]), and papers by S.D. Webb in *Paleobiology* (vol. 2, pp. 220–34 [1976] and Vol. 17, pp. 266–80 [1991]).

6. That at least is a widely held notion. Nevertheless, there may now be serious reasons to doubt the hypothesis of extinction by competition after the interchange. Lessa and Farina in *Palaeontology* (Vol. 39, pp. 651–62 [1996]) have questioned the proposal that the demise of the South American mammalian fauna was due to competitive interactions with the northern invaders. They point out that the pattern of extinction is strongly correlated to large body size, and suggest that the disappearances may be linked either to climate change or to human

hunting, both occurring long after the land bridge between North and South America was established. These workers also point out that the differential success enjoyed by the northern mammals is apparently due to their greater rates of biological diversification.

7. See his chapter in *Rates of evolution* (ed. K.S.W. Campbell and M.F. Day), pp. 39–60 (Allen and Unwin, London, 1987), where he gives a useful overview with respect to the molluscs, and also acknowledges the formative influence of Valdar Jannusson in the context of articulating the difference between disparity and diversity.

8. Palaeontologists are well aware of the phenomenon, articulated by Dave Raup, known as the 'Pull of the Recent'. In essence this concept warns researchers that as one approaches the present day so the rock record becomes more complete and less subject to diagenesis and metamorphism. In addition, the containing biotas are increasingly similar to assemblages still living, and this allows more detailed comparisons to be made. In short, can we be sure that the observed climb in diversity towards the Recent is not an artefact of sampling? Removing the biases in an incomplete stratigraphic record so as to reveal the actual history of diversity is not straightforward, but overall the evidence that the last few million years are the most diverse time ever seen in the history of the planet seems credible. Not only that but study of complex ecosystems, such as coral reefs, also suggest a level of sophistication unseen in the Palaeozoic and Mesozoic. The irony of Man arriving on such a fecund scene and promptly acting to degrade it to a depauperate shadow of its former self deserves to be more widely appreciated.

9. Moulting in trilobites has been studied by a number of workers. The paper by K.J. McNamara and D.M. Rudkin in *Lethaia* (Vol. 17, pp. 153–7 [1984]) is particularly useful.

10. The paper describing this analysis of trilobite disparity in terms of cephalon shape was published in *Palaeontology* (Vol. 34, pp. 461–85 [1991]).

11. A very interesting analysis of trilobite decline was published by D.M. Raup in *Acta Geologica Hispanica* (Vol. 16, pp. 25–33 [1981]), in which he concluded that the trilobites did not simply drift to extinction, but failed against other organisms. The nature of this failure ('bad genes') as against being in the wrong place at the wrong time ('bad luck') is open to discussion, although suspicion may focus on competitive interactions with other marine animals.

12. The details of this technique are given by M. Foote in *Journal of Paleontology* (Vol. 63, pp. 880–5 [1989]).

13. This paper, which also considered aspects of disparity in other fossil groups, was published in *Paleobiology* (Vol. 19, pp. 185–204 [1993]). Subsequently with A.I. Miller, Mike Foote published a more detailed analysis of changes in the disparity of Ordovician trilobites (*Paleobiology*, vol. 22, pp. 304–9 [1996]). This demonstrated the first half of the Ordovician period was marked by a steep climb in disparity, whereas the remainder displayed a plateau.

14. See his paper in *Systematic Zoology* (Vol. 39, pp. 371–82 [1990]).

15. See notes 10 and 12 of this chapter. The analysis by D.M. Raup looks at the trilobites as a whole, whereas that by M. Foote considered several major groups whose behaviour differs in various ways. In a group known as the Scutelluina, for example, there appears to be good evidence for selective extinction of particular types that is therefore reflected in their changing history of disparity.

16. See their paper in *Paleobiology* (Vol. 20, pp. 93–130 [1994]).

17. Even so, one must observe that length of list may not be directly equated with rigour of argument. The tendency in such analyses, as in cladistic formulations, for an atomistic dissection of organisms into a myriad of character states raises two problems. First, the extent to which characters are intercorrelated, so that a large data set actually contains a large measure of redundancy. In other words, if one character changes then others automatically follow suit. There is, of course, a close link to the genetics of developmental biology whereby the hierarchy of gene

expression, if perturbed or otherwise altered, may lead to a series of correlated changes. Second, the relative weight (if any) to be given to each character. For some characters there may be a near-consensus that they are indeed fundamental. Others may be so regarded, but in fact transpire to be evolutionarily labile. Finally, others may be rather trivial, whose contribution to perceived disparity is exaggerated.

18. A critique by Mike Foote, Steve Gould, and Mike Lee was published in *Science* (Vol. 258, pp. 1816–17 [1992]), and a reply offered by Derek Briggs and others (pp. 1817–18).

19. The measurements of disparity in various groups of invertebrate fossils and their comparison with taxonomic richness are reviewed by M. Foote in a chapter of *Evolutionary paleobiology, in honor of James W. Valentine* (ed. D. Jablonski, D.H. Erwin, and J.H. Lipps), pp. 62–86 (University of Chicago Press, 1996). A comparable excursion into vertebrates, specifically Tertiary ungulates, is given by J. Jervall and others (*Science*, Vol. 274, pp. 1489–92 [1996]). In their analysis of the disparity of molar teeth, a key element in ungulate existence, they found increasing disparity through geological time. Interestingly, these workers suggested that changes in disparity could be linked to adaptive explanations, and that convergent evolution was frequent. Both these points are discussed at some length elsewhere in this book.

CHAPTER 9

Last word

One of the unshaken orthodoxies of science is that the Galaxy, and by implication much of the visible Universe, teems with life. Most probably it does.
There is not quite the same degree of consensus, but if asked most scientists
would probably agree that any one time the Galaxy houses advanced civilizations. Most estimates suggest that there are several such civilizations, perhaps
tens, maybe even hundreds. Scientific fascination with these speculations
largely centres on the technological achievements of alien societies, and there
is little concern about their moral and spiritual status. Our attempts to make
contact with an extraterrestrial civilization have only just begun. All one can
say is that, so far, it seems to be remarkably quiet out there. There is also no
evidence, at least that we have recognized, that the Earth has ever been
visited by aliens. It is probably optimistic to imagine that any traces of such a
visit would survive, especially if it occurred millions of years ago. In addition,
for all we know one clause in the *Galactic Code for Visitors* insists that
visitors leave no evidence of their stay.

But let us suppose that not only civilizations, but even life itself is unique to
the Earth.[1] At first sight this seems to be a ridiculous statement. Perhaps it is.
There is no questioning the ease of constructing the basic building blocks of
life, such as the amino acids, carbohydrates, and lipids. For many years, labo-
ratories all over the world have been investigating these problems, and have
achieved many successful syntheses. It is also clear that a wide variety of
simple organic compounds occur in deep space and may even reach the Earth
in meteorites of a special type (the carbonaceous meteorites). But suppose
that the far more complex phenomenon of life itself, defined as a replicating
cell, was the product of a freak and extraordinarily rare event. Imagine also
that in fact relatively few stars have planets, at least of a suitable size and
composition for habitation. Of those that do, some will be so hot that the
zone of habitation will be on the edge of the planetary system, with most of
its planets roasting in a torrid zone. Other stars may be so dim that they are
orbited by frozen globes. In either case the planetary systems will be dead, the
planets nothing more than sterile balls of rock.

Such speculations are not as wild as it is sometimes believed. Synthesizing
the chemical building blocks of life is no guarantee that life itself is a pre-
ordained inevitability. Even if life is eventually created in the laboratory, this
will be by conscious manipulation. What happened on the Earth four billion
years ago may have been very different. Planets circling distant stars have

already been detected, but in itself this tells us almost nothing about the probability of alien life existing.

Perhaps as you read these pages the radio will be filled with news of the first extraterrestrial contact. But the alternative proposition, that we are alone, is worth thinking about seriously. Let us accept, for the sake of the argument, that not only is life unique to the Earth, but that our species is without parallel either with the millions of other species with which we share our planet or the billions, now extinct, that preceded us. It is the opinion of many individuals that our present behaviour in terms of the abuse of ecological systems, the continuing degradation of the land and seas, and the profligacy of our consumption are all severely deleterious. The present trends cannot be sustained either in terms of preserving the richness of biodiversity or for our own spiritual health. Until we act as responsible stewards of our planet and stop behaving like unwelcome guests, it is likely that we will enter a period of worsening crises, especially in terms of environmentally induced disasters and urban trauma.

All these points are being made forcibly by people far more articulate on these matters than I am. I should like to conclude on a different note, albeit related. If indeed we are alone and unique, and this possibility, however implausible, cannot yet be refuted, then we have special responsibilities. First, as alluded to above, our present behaviour is little short of reckless. Second, it follows that we have special duties to our descendants, at least to leave the world a little better than we found it. But there is also a unique privilege. That is to understand a little of our history. For some it will be the investigation of the past few centuries; for others the search for the origins of mankind. Yet others will wish to reach further back in geological time to discover in the Cambrian period the seeds of our own destiny. This quest started with Walcott's great discovery of the Burgess Shale, and continues to gain momentum with the new discoveries in China and Greenland. Much has been learnt, but by no means the last page of this story has been written.

Note on Chapter 9

1. This topic deserves a book to itself, but the following points are ones that can contribute to the debate. In particular, the following factors may seriously constrain the likelihood of finding life elsewhere. (1) The centres of galaxies may be occupied by black holes, rendering most galactic cores uninhabitable. (2) Many star systems are binary, and the complexities of planetary orbits, while making for spectacular sunsets, may preclude life. M. Holman and others (*Nature*, Vol. 386, pp. 254–6 [1997]) have, for example, argued that a planet orbiting the Star 16 Cyg B has a wildly eccentric orbit that is probably induced by a companion star (16 Cyg A), and that such eccentricity may lead to the planet plunging into the star it orbits. (3) Evidence grows for planets orbiting other planets, although recent results stemming from a re-analysis of spectral line shapes of one star has cast severe doubt on one such interpretation (see D.F. Gray's paper in

Nature, Vol. 385, pp. 795–6 [1997]). Remember also that eight of the nine planets in our Solar System are as dead as doornails; this too should give us pause for thought. More importantly, the calculations of planetary sizes and distances from the Sun give a very wide range of possibilities, few of which may either fall into the habitable zone or be small enough to allow a manageable gravitational field. (4) Calculations of cometary orbits indicate that practically all those observed derive from the Oort Clouds and seldom, if ever, do they have a hyperbolic orbit which would indicate derivation from interstellar space and by implication another solar system (see the book by J.C. Brandt and R.D. Chapman entitled *Rendezvous in space: the science of comets* (Freeman, New York, 1992)). As these authors point out this is decidedly puzzling, because the comet population of our Solar System is estimated to run into billions, and calculations suggest that many comets routinely escape into interstellar space. As Brandt and Chapman note (p. 170) 'Perhaps solar systems like ours are the exception rather than the rule'. There are, of course, other explanations for the apparent absence of interstellar comets visiting us, and Brandt and Chapman suggest some ingenious observations that might be made in an attempt to detect equivalent Oort Clouds around other stars. (5) Our large daughter satellite, the Moon, may be very unusual. In addition to maintaining axial stability of the Earth it has also been an important ingredient in allowing the surface of the Earth to be habitable. In this context see *What if the Moon didn't exist?* by N.F. Comins (HarperPerennial, New York, 1993). It might be still argued that the number of potential habitable worlds remains large. Now let us extend the arguments on contingency employed so forcibly by Steve Gould and enquire why the origination of life should be immune to the rerunning of his metaphorical tape? After all, on this basis it could be argued with equal force that in every other planet the 'warm, little pond' remained exactly that, until it either dried out or froze over.

Whether the foregoing summary of evidence, which may drastically narrow the likelihood of finding life beyond the Solar System, wins any acceptance remains to be seen. Of yet greater moment is the possibility of contacting anything akin to human intelligence. Given the ubiquity of opinion that intelligent extraterrestrial life is at the least highly probable, it is difficult to find countervailing voices. One such, of considerable articulation and force, is that of S.L. Jaki. See especially his book *God and the cosmologists* (Scottish Academic Press, Edinburgh, 1989), and also *Is there a Universe?* (Liverpool University Press, 1993).

Appendix 1. Further reading

For those who wish to explore other aspects of the Burgess Shale, I list here some of the popular books, together with some brief comments. Following this is a short list of scientific literature which will provide an introduction to this subject at a more advanced level.

Popular books

Briggs, D.E.G., Erwin, D.H., and Collier, F.J. (1994). *The fossils of the Burgess Shale*. Smithsonian Institution Press, Washington. Copiously illustrated, although not all the photographs really do justice to the fossils. The explanatory text is straightforward, but there is no real attempt to address current controversies or the wider significance of the Burgess Shale.

Gould, S.J. (1989). *Wonderful life. The Burgess Shale and the nature of the history*. Norton, New York The first book to draw wide public attention to the Burgess Shale, largely describing the research in Cambridge.

Whittington, H.B. (1985). *The Burgess Shale*. Yale University Press, New Haven. A straightforward and concise description of the Burgess Shale fauna by the leader of the Cambridge team.

In my opinion there is no really good book by a single author on the early evolution of animals and the Cambrian 'explosion'. The best on the market is still by M.F. Glaessner (1984). *The dawn of animal life. A biohistorical study* (Cambridge University Press), but it is becoming quite dated. Two multi-author volumes can also be recommended: Bengtson, S. (ed.) (1994). *Early life on Earth* (Columbia University Press); and Lipps, J.H. and Signor, P.W. (eds.) (1992). *Origin and early evolution of the Metazoa* (Plenum Press, New York). The huge volume edited by J.W. Schopf and C. Klein (1992) *The Proterozoic biosphere: a multidisciplinary study*. (Cambridge University Press) provides a series of relevant vignettes, but there is rather little critical discussion.

Interest in the Burgess Shale has also extended to popular fiction. Two examples include the books by William Gibson and Bruce Sterling: *The difference engine* (1990, Victor Gollancz, London) and Penelope Lively: *Cleopatra's Sister* (1993, Viking, New York) respectively.

Scientific papers

There is substantial literature on the Burgess Shale and similar faunas and their role in the Cambrian 'explosion'; the end-chapter notes make reference to a number of pertinent examples. Because of the scattered nature of these references it was thought sensible to bring most of these, and a number of others, into a single list which, although making no pretensions to completeness, does aim to help those readers who wish to explore the more technical literature.

Aitken, J.D., and McIlreath, I.A. (1984). The Cathedral Reef Escarpment, a Cambrian great wall with humble origins. *Geos*, **13**, 17–19.

Aitken, J.D., and McIlreath, I.A. (1990). Comment [In defense of the Escarpment near the Burgess Shale fossil locality]. *Geoscience Canada* 17, 111–16.

Allison, P.A., and Brett, C.E. (1995). In situ benthos and paleo-oxygenation in the Middle Cambrian Burgess Shale, British Columbia, Canada. *Geology*, 23, 1079–82.

Aronson, R.B. (1992). Decline of the Burgess Shale fauna: ecologic or taphonomic restriction? *Lethaia*, 25, 225–9.

Bergström, J. (1986). *Opabinia* and *Anomalocaris*, unique Cambrian 'arthropods'. *Lethaia*, 19, 241–6.

Blaker, M.R. (1988). A new genus of nevadiid trilobite from the Buen Formation (Early Cambrian) of Peary Land, central north Greenland. *Rapport Grønlands Geologiske Undersøgelse*, 137, 33–41.

Bousfield, E.L. (1995). A contribution to the natural classification of Lower and Middle Cambrian arthropods: food-gathering and feeding mechanisms. *Amphipacifica*, 2, 3–34.

Briggs, D.E.G. (1976). The arthropod *Branchiocaris* n.gen., Middle Cambrian, Burgess Shale, British Columbia. *Bulletin of the Geological Survey of Canada*, 264, 1–29.

Briggs, D.E.G. (1978). A new trilobite-like arthropod from the Lower Cambrian Kinzers Formation, Pennsylvania. *Journal of Paleontology*, 52, 132–40.

Briggs, D.E.G. (1978). The morphology, mode of life, and affinities of *Canadaspis perfecta* (Crustacea: Phyllocarida), Middle Cambrian, Burgess Shale, British Columbia. *Philosophical Transactions of the Royal Society, London*, B 281, 439–87.

Briggs, D.E.G. (1979). *Anomalocaris*, the largest known Cambrian arthropod. *Palaeontology*, 22, 631–64.

Briggs, D.E.G. (1981). The arthropod *Odaraia alata* Walcott, Middle Cambrian, Burgess Shale, British Shale. *Philosophical Transactions of the Royal Society, London*, B 291, 541–84.

Briggs, D.E.G. (1992). Phylogenetic significance of the Burgess Shale crustacean *Canadaspis*. *Acta Zoologica (Stockholm)*, 73, 293–300.

Briggs, D.E.G. and Collins, D. (1988). A Middle Cambrian chelicerate from Mount Stephen, British Columbia. *Palaeontology*, 31, 779–98.

Briggs, D.E.G. and Fortey, R.A. (1989). The early radiation and relationships of the major arthropod groups. *Science*, 246, 241–3.

Briggs, D.E.G. and Mount, J.D. (1982). The occurrence of the giant arthropod *Anomalocaris* in the Lower Cambrian of southern California, and the overall distribution of the genus. *Journal of Paleontology*, 56, 1112–18.

Briggs, D.E.G. and Nedin, C. (1997). The taphonomy and affinities of the problematic fossil *Myoscolex* from the Lower Cambrian Emu Bay Shale of South Australia. *Journal of Paleontology*, 71, 22–32.

Briggs, D.E.G. and Robison, R.A. (1984). Exceptionally preserved nontrilobite arthropods and *Anomalocaris* from the Middle Cambrian of Utah. *University of Kansas Paleontological Contribution*, 111, 1–23.

Briggs, D.E.G. and Whittington, H.B. (1985). Modes of life of arthropods from the Burgess Shale, British Columbia. *Philosophical Transactions of the Royal Society of Edinburgh*, 76, 149–60.

Bruton, D. (1981). The arthropod *Sidneyia inexpectans*, Middle Cambrian, Burgess Shale, British Columbia. *Philosophical Transactions of the Royal Society of London*, B 295, 619–56.

Bruton, D.L., Jensen, A., and Jacquet, R. (1985). The use of models in the understanding of Cambrian arthropod morphology. *Transactions of the Royal Society of Edinburgh*, 76, 365–9.

Bruton, D.L. and Whittington, H.B. (1983). *Emeraldella* and *Leanchoilia*, two arthropods from the Burgess Shale, Middle Cambrian, British Columbia. *Philosophical Transactions of the Royal Society of London*, B 300, 553–85.

Budd, G. (1993). A Cambrian gilled lobopod from Greenland. *Nature*, 364, 709–11.

Budd, G.E. (1995). *Kleptothule rasmusseni* gen. et sp. nov.: an ?olenellid-like trilobite from the Sirius Passet fauna (Buen Formation, Lower Cambrian, North Greenland). *Transactions of the Royal Society of Edinburgh: Earth Sciences*, 86, 1–12.

Budd, G.E. (1996). The morphology of *Opabinia regalis* and the reconstruction of the arthropod stem-group. *Lethaia*, 29, 1–14.

Butterfield, N.J. (1990). Organic preservation of non-mineralizing organisms and the taphonomy of the Burgess Shale. *Paleobiology*, 16, 272–86.

Butterfield, N.J. (1990). A reassessment of the enigmatic Burgess Shale fossil *Wiwaxia corrugata* (Matthew) and its relationship to the polychaete *Canadia spinosa* Walcott. *Paleobiology*, 16, 287–303.

Butterfield, N.J. (1994). Burgess Shale-type fossils from a Lower Cambrian shallow-shelf sequence in northwestern Canada. *Nature*, 369, 477–9.

Butterfield, N.J. (1995). Secular distribution of Burgess Shale-type preservation. *Lethaia*, 28, 1–13.

Butterfield, N.J. (1996). Fossil preservation in the Burgess Shale: Reply. *Lethaia*, 29, 109–12.

Butterfielsd, N.J. (1997). Plankton ecology and the Proterozoic–Phanerozoic transition. *Paleobiology*, 23, 247–62.

Butterfield, N.J. and Nicholas, C.J. (1996). Burgess Shale-type preservation of both non-mineralizing and 'shelly' Cambrian organisms from the Mackenzie Mountains, northwestern Canada. *Journal of Paleontology*, 70, 893–9.

Campbell, L.D. (1971). Occurrence of "*Ogygopsis* shale" fauna in southeastern Pennsylvania. *Journal of Paleontology*, 45, 437–40.

Chen Junyuan, Edgecombe, G.D., and Ramsköld, L. (1997). Morphological and ecological disparity in naraoiids (Arthropoda) from the Early Cambrian Chengjiang fauna, China. *Records of the Australian Museum*, 49, 1–24.

Chen Junyuan, Edgecombe, G.D., Ramsköld, L., and Zhou Guiqing. (1995). Head segmentation in early Cambrian *Fuxianhuia*: Implications for arthropod evolution. *Science*, 268, 1339–43.

Chen Junyuan, Ramsköld, L., and Zhou Guiqing. (1994). Evidence for monophyly and arthropod affinity of Cambrian predators. *Science*, 264, 1304–8.

Chen Junyuan, Zhou Guiqing, and Ramsköld, L. (1995). A new Early Cambrian onychophoran-like animal *Paucipodia* gen. nov., from the Chengjiang fauna, China. *Transactions of the Royal Society of Edinburgh: Earth Sciences*, 85, 275–82.

Chen Junyuan, Zhou Guiqing, Zhu Maoyan, and Yeh K.Y. (ca. 1996). *The Chengjiang biota. A unique window of the Cambrian explosion*. National Museum of Natural Science, Taiwan. [In Chinese]

Chen Junyuan, Zhu Maoyan, and Zhou Guiqing. (1995). The early Cambrian medusiform metazoan *Eldonia* from the Chengjiang Lagerstätte. *Acta Palaeontologica Polonica*, 40, 213–44.

Collins, D. (1996). The 'evolution' of *Anomalocaris* and its classification in the arthropods Class Dinocarida (Nov.) and Order Radiodonta (Nov.). *Journal of Paleontology*, 70, 280–93.

Collins, D., Briggs, D., and Conway Morris, S. (1983). New Burgess Shale fossil sites reveal Middle Cambrian faunal complex. *Science*, 222, 163–7.

Collins, D. and Rudkin, D.M. (1981). *Priscansermarinus barnetti*, a probable lepadomorph barnacle from the Middle Cambrian Burgess Shale of British Columbia. *Journal of Paleontology*, 55, 1006–15.

Conway Morris, S. (1976). A new Cambrian lophophorate from the Burgess Shale of British Columbia. *Palaeontology*, 19, 199–222.

Conway Morris, S. (1976). *Nectocaris pteryx*, a new organism from the Middle Cambrian Burgess Shale of British Columbia. *Neues Jahrbuch für Geologie und Paläontologie*, Monatshefte H12, 705–13.

Conway Morris, S. (1977). A new metazoan from the Cambrian Burgess Shale of British Columbia. *Palaeontology*, 20, 623–40.

Conway Morris, S. (1977). A new entoproct-like organism from the Burgess Shale of British Columbia. *Palaeontology*, **20**, 833–45.

Conway Morris, S. (1977). Fossil priapulid worms. *Special Papers in Palaeontology*, **20**, i–iv, 1–95.

Conway Morris, S. (1977). A redescription of the Middle Cambrian worm *Amiskwia sagittiformis* Walcott from the Burgess Shale of British Columbia. *Paläontologische Zeitschrift*, **51**, 271–87.

Conway Morris, S. (1979). The Burgess Shale (Middle Cambrian) fauna. *Annual Review of Ecology and Systematics*, **10**, 327–49.

Conway Morris, S. (1979). Middle Cambrian polychaetes from the Burgess Shale of British Columbia. *Philosophical Transactions of the Royal Society of London*, B **285**, 227–74.

Conway Morris, S. (ed.) (1982). *Atlas of the Burgess Shale*. Palaeontological Association, London.

Conway Morris, S. (1985). The Middle Cambrian metazoan *Wiwaxia corrugata* (Matthew) from the Burgess Shale and *Ogygopsis* Shale, British Columbia, Canada. *Philosophical Transactions of the Royal Society of London*, B **307**, 507–86.

Conway Morris, S. (1985). Cambrian Lagerstätten: their distribution and significance. *Philosophical Transactions of the Royal Society of London*, B **311**, 49–65.

Conway Morris, S. (1986). The community structure of the Middle Cambrian Phyllopod bed (Burgess Shale). *Palaeontology*, **29**, 423–67.

Conway Morris, S. (1989). Burgess Shale faunas and the Cambrian explosion. *Science*, **246**, 339–46.

Conway Morris, S. (1989). The persistence of Burgess Shale-type faunas: implications for the evolution of deeper-water faunas. *Transactions of the Royal Society of Edinburgh: Earth Sciences*, **80**, 271–83.

Conway Morris, S. (1990). Late Precambrian and Cambrian soft-bodied faunas. *Annual Review of Earth and Planetary Sciences*, **18**, 101–22.

Conway Morris, S. (1992). Burgess Shale-type faunas in the context of the 'Cambrian explosion': a review. *Journal of the Geological Society, London*, **149**, 631–6.

Conway Morris, S. (1993). Ediacaran-like fossils in Cambrian Burgess Shale-type faunas of North America. *Palaeontology*, **36**, 593–635.

Conway Morris, S. (1993). The fossil record and the early evolution of the Metazoa. *Nature*, **361**, 219–25.

Conway Morris, S. (1994). Why molecular biology needs palaeontology. *Development, Supplement* (for 1994), 1–13.

Conway Morris, S. (1995). Enigmatic shells, possibly halkieriid, from the Middle Cambrian Burgess Shale, British Columbia. Neues *Jahrbuch für Geologie und Paläontologie, Abhandlungen*, **195**, 319–31.

Conway Morris, S. and Peel, J.S. (1995). Articulated halkieriids from the Lower Cambrian of North Greenland and their role in early protostome evolution. *Philosophical Transactions of the Royal Society of London*, B **347**, 305–58.

Conway Morris, S. and Robison, R.A. (1982). The enigmatic medusoid *Peytoia* and a comparison of some Cambrian biotas. *Journal of Paleontology*, **56**, 116–22.

Conway Morris, S. and Robison, R.A. (1986). Middle Cambrian priapulids and other soft-bodied fossils from Utah and Spain. *University of Kansas Paleontological Contributions*, **117**, 1–22.

Conway Morris, S. and Robison, R.A. (1988). More soft-bodied animals and algae from the Middle Cambrian of Utah and British Columbia. *University of Kansas Paleontological Contributions*, **122**, 1–48.

Conway Morris, S. and Whittington, H.B. (1985). Fossils of the Burgess Shale, a national treasure in Yoho National Park, British Columbia. *Miscellaneous Reports of the Geological Survey of Canada*, **43**, 1–31.

Durham, J.W. (1974). Systematic position of *Eldonia ludwigi* Walcott. *Journal of Paleontology*, **48**, 750–5.

Dzik, J. (1995). *Yunnanozoon* and the ancestry of chordates. *Acta Palaeontologica Polonica*, 40, 341–60.

Dzik, J., Zhao Yuanlong, and Zhu Maoyan. (1997). Mode of life of the Middle Cambrian eldonioid lophophrate *Rotadiscus*. *Palaeontology*, 40, 385–96.

Erwin, D.M. (1993). The origin of metazoan development: a palaeobiological perspective. *Biological Journal of the Linnean Society*, 50, 255–74.

Fritz, W.H. (1971). Geological setting of the Burgess Shale. In *Symposium on Extraordinary Fossils. Proceedings of the North American Paleontological Convention, Field Museum of Natural History, Chicago. September 5–7, 1969*, Part I, 1155–70. Allen Press, Lawrence, Kansas.

Fritz, W.H. (1990). Comment [In defense of the Escarpment near the Burgess Shale fossil locality]. *Geoscience Canada*, 17, 106–10.

Glaessner, M.F. (1979). Lower Cambrian Crustacea and annelid worms from Kangaroo Island, South Australia. *Alcheringa*, 3, 21–31.

Gould, S.J. (1991). The disparity of the Burgess Shale arthropod fauna and the limits of cladistic analysis: why we must strive to quantify morphospace. *Paleobiology*, 17, 411–23.

Grotzinger, J.P., Bowring, S.A., Saylor, B.Z., and Kaufman, A.J. (1995). Biostratigraphic and geochronologic constraints on early animal evolution. *Science*, 270, 598–604.

Hou Xianguang and Bergström, J. (1994). Palaeoscolecid worms may be nematomorphs rather than annelids. *Lethaia*, 27, 11–17.

Hou Xianguang and Bergström, J. (1995). Cambrian lobopodians—ancestors of extant onychophorans? *Zoological Journal of the Linnean Society*, 114, 3–19.

Hou Xianguang, Bergström, J., and Ahlberg, P. (1995). *Anomalocaris* and other large animals in the Lower Cambrian of southwest China. *Geologiska Föreningens i Stockholm Förhandlingar*, 117, 163–83.

Hou Xianguang, Ramsköld, L., and Bergström, J. (1991). Composition and preservation of the Chengjiang fauna—a Lower Cambrian soft-bodied biota. *Zoologica Scripta*, 20, 395–411.

Hou Xianguang, Siveter, D.J., Williams, M., Walossek, D., and Bergström, J. (1996). Appendages of the arthropod *Kunmingella* from the early Cambrian of China: its bearing on the systematic position of the Bradoriida and the fossil record of the Ostracoda. *Philosophical Transactions of the Royal Society of London*, B 351, 1131–45.

Hughes, C.P. (1975). Redescription of *Burgessia bella* from the Middle Cambrian Burgess Shale, British Columbia. *Fossils and Strata*, 4, 415–35.

Jin Yugan, Hou Xianguang, and Wang Huayu. (1993). Lower Cambrian pediculate lingulids from Yunnan, China. *Journal of Paleontology*, 67, 788–98.

Jin Yugan and Wang Huayu. (1992). Revision of the Lower Cambrian brachiopod *Heliomedusa* Sun and Hou, 1987. *Lethaia*, 25, 35–49.

Ludvigsen, R. (1989). The Burgess Shale: Not in the shadow of the Cathedral Escarpment. *Geoscience Canada*, 16, 51–9.

Ludvigsen, R. (1990). Reply to comments by Fritz and Aitken and McIlreath. *Geoscience Canada*, 17, 116–18.

Luo Huilin, Hu Shixue, Zhang shishan and Tao Yonghe. (1997). New occurrence of the early Cambrian Chengjiang fauna from Haikou, Kunming, Yunnan province, and study of Trilobitoidea. *Acta Geologica Sinica*, 71, 97–104.

McHenry, B., and Yates, A. (1993). First report of the enigmatic metazoan *Anomalocaris* from the southern hemisphere and a trilobite with preserved appendages from the early Cambrian of Kangaroo Island, South Australia. *Records of the South Australian Museum*, 26, 77–86.

McIlreath, I.A. (1977). Accumulation of a Middle Cambrian, deep-water limestone debris apron adjacent to a vertical, submarine carbonate escarpment, southern Rocky Mountains, Canada. *Society of Economic Paleontologists and Mineralogists Special Publication*, 25, 113–24.

Mankiewicz, C. (1992). *Obruchevella* and other microfossils in the Burgess Shale: preservation and affinity. *Journal of Paleontology*, **66**, 717–29.

Masiak, M. and Zylinska, A. (1994). Burgess Shale-type fossils in Cambrian sandstones of the Holy Cross Mountains. *Acta Palaeontologica Polonica*, **39**, 329–40.

Nedin, C. (1995). The Emu Bay Shale, a Lower Cambrian fossil Lagerstätten [sic], Kangaroo Island, South Australia. *Memoirs of the Association of Australasian Palaeontologists*, **18**, 31–40.

Piper, D.J.W. (1972). Sediments of the Middle Cambrian Burgess Shale, Canada. *Lethaia*, **5**, 169–75.

Ramsköld, L. (1992). Homologies in Cambrian Onychophora. *Lethaia*, **25**, 443–60.

Ramsköld, L., Chen Junyuan, Edgecombe, G.D., and Zhou Guiqing. (1996). Preservational folds simulating tergite junctions in tegopeltid and naraoiid arthropods. *Lethaia*, **29**, 15–20.

Ramsköld, L., Chen Junyuan, Edgecombe, G.D., and Zhou Guiqing. (1997). *Cindarella* and the arachnate clade Xandarellida (Arthropoda, Early Cambrian) from China. *Transactions of the Royal Society of Edinburgh: Earth Sciences*, **88**, 19–38.

Ramsköld, L. and Hou Xianguang. (1991). New early Cambrian animal and onychophoran affinities of enigmatic metazoans. *Nature*, **351**, 225–8.

Rigby, J.K. (1980). The new Middle Cambrian sponge *Vauxia magna* from the Spence Shale of northern Utah and taxonomic position of the Vauxiidae. *Journal of Paleontology*, **54**, 234–40.

Rigby, J.K. (1983). Sponges of the Middle Cambrian Marjum Limestone from the House Range and Drum Mountains of western Millard County, Utah. *Journal of Paleontology*, **57**, 240–70.

Rigby, J.K. (1986). Sponges of the Burgess Shale (Middle Cambrian), British Columbia. *Palaeontographica Canadiana*, **2**, 1–105.

Robison, R.A. and Wiley, E.O. (1995). A new arthropod, *Meristosoma*: More fallout from the Cambrian explosion. *Journal of Paleontology*, **69**, 447–59.

Shu Degan, Conway Morris, S., and Zhang Xianliang. (1996). A *Pikaia*-like chordate from the Lower Cambrian of China. *Nature*, **384**, 157–8.

Shu Degan, Geyer, G., Chen Ling, and Zhang Xingliang. (1995). Redlichiacean trilobites with preserved soft-parts from the Lower Cambrian Chengjiang fauna (South China). In *Morocco '95. The Lower Middle Cambrian standard of western Gondwana* (ed. G. Geyer and E. Landing), pp. 203–40. *Beringeria, Special Issue 2*.

Shu Degan, Zhang Xingliang, and Cheng Ling. (1996). Reinterpretation of *Yunnanozoon* as the earliest known hemichordate. *Nature*, **380**, 428–30.

Shu Degan, Zhang Xingliang, and Geyer, G. (1995). Anatomy and systematic affinities of the Lower Cambrian bivalved arthropod *Isoxys auritus*. *Alcheringa*, **19**, 333–42.

Simonetta, A.M. (1988). Is *Nectocaris pteryx* a chordate? *Bollettino Zoologica*, **55**, 63–8.

Simonetta, A.M. and Conway Morris, S. (eds). (1991). *The early evolution of Metazoa and the significance of problematic taxa*. Cambridge University Press.

Simonetta, A.M. and Insom, E. (1993). New animals from the Burgess Shale (Middle Cambrian) and their possible significance for the understanding of the Bilateria. *Bollettino Zoologica*, **60**, 97–107.

Towe, K.M. (1996). Fossil preservation in the Burgess Shale. *Lethaia*, **29**, 107–8.

Whittington, H.B. (1971). The Burgess Shale: history of research and preservation of fossils. In *Symposium on extraordinary fossils. Proceedings of the North American Paleontological Convention, Field Museum of Natural History, Chicago, September 5–7, 1969*, Part I, 1170–201. Allen Press, Lawrence, Kansas.

Whittington, H.B. (1971). Redescription of *Marrella splendens* (Trilobitoidea) from the Burgess Shale, Middle Cambrian, British Columbia. *Bulletin of the Geological Survey of Canada*, **209**, 1–24.

Whittington, H.B. (1974). *Yohoia* Walcott and *Plenocaris* n. gen., arthropods from the Burgess Shale, Middle Cambrian, British Columbia. *Bulletin of the Geological Survey of Canada*, 231, 1–27.

Whittington, H.B. (1975). The enigmatic animal *Opabinia regalis*, Middle Cambrian, Burgess Shale, British Columbia. *Philosophical Transactions of the Royal Society of London*, B 271, 1–43.

Whittington, H.B. (1975). Trilobites with appendages from the Middle Cambrian, Burgess Shale, British Columbia. *Fossils and Strata*, 4, 97–136.

Whittington, H.B. (1977). The Middle Cambrian trilobite *Naraoia*, Burgess Shale, British Columbia. *Philosophical Transactions of the Royal Society of London*, B 280, 409–43.

Whittington, H.B. (1978). The lobopod animal *Aysheaia pedunculata*, Middle Cambrian, Burgess Shale, British Columbia. *Philosophical Transactions of the Royal Society of London*, B 284, 165–97.

Whittington, H.B. (1979). Early arthropods, their appendages and relationships. In *The origin of major invertebrate groups* (ed. M.R. House). Systematics Association Special Volume 12, 253–68.

Whittington, H.B. (1980). The significance of the fauna of the Burgess Shale, Middle Cambrian, British Columbia. *Proceedings of the Geological Association*, 91, 127–48.

Whittington, H.B. (1980). Exoskeleton, moult stage, appendage morphology, and habits of the Middle Cambrian trilobites *Olenoides serratus*. *Palaeontology*, 23, 171–204.

Whittington, H.B. (1981). Rare arthropods from the Burgess Shale, Middle Cambrian, British Columbia. *Philosophical Transactions of the Royal Society of London*, B 292, 329–57.

Whittington, H.B. (1985). *Tegopelte gigas*, a second soft-bodied trilobite from the Burgess Shale, Middle Cambrian, British Columbia. *Journal of Paleontology*, 59, 1251–74.

Whittington, H.B. and Briggs, D.E.G. (1985). The largest Cambrian animal, *Anomalocaris*, Burgess Shale, British Columbia. *Philosophical Transactions of the Royal Society of London*, B 309, 569–609.

Williams, M., Siveter, D.J., and Peel, J.S. (1996). *Isoxys* (Arthropoda) from the early Cambrian Sirius Passet Lagerstätte, North Greenland. *Journal of Paleontology*, 70, 947–54.

Wills, M.A., Briggs, D.E.G., and Fortey, R.A. (1994). Disparity as an evolutionary index: a comparison of Cambrian and Recent arthropods. *Paleobiology*, 20, 93–130.

Yochelson, E.L. (1996). Discovery, collection, and description of the Middle Cambrian Burgess Shale biota by Charles Doolittle Walcott. *Proceedings of the American Philosophical Society*, 140, 469–545.

Zhao Yuanlong *et al.* (1994). Middle Cambrian Kaili fauna in Taijiang, Guizhou. *Acta Palaeontologica Sinica*, 33, 269–71. [In Chinese, with English abstract].

Appendix 2. Exhibitions

Charles Walcott (and his assistant Charles Resser) distributed small collections of Burgess Shale fossils to many institutions, but the great bulk reside in the Smithsonian Institution (Washington, DC) and the Royal Ontario Museum (Toronto). The following museums have some sort of display of the Burgess Shale.

1. National Museum of Natural History, Smithsonian Institution, Washington DC, USA. At the time of writing the existing display is due to be replaced with a new exhibit.
2. Royal Ontario Museum, Toronto, Canada.
3. Field Visitor Centre, Field, British Columbia, Canada. Displays are also at Monarch Campground on Yoho Valley Road (i.e. the road near Field leading to the Takakkaw Falls) and Lake Louise Visitor Centre.
4. Royal Tyrell Museum, Drumheller, Alberta, Canada.
5. Sedgwick Museum, University of Cambridge, Cambridge, England.

Appendix 3. Localities

1. The Burgess Shale, Canada. This is located in Yoho National Park and visits to the actual quarries are strictly controlled. This applies also to Mount Stephen. Licensed private guides provide trips to both Walcott's quarry and Mount Stephen. Each trip is limited to 15 people, and must be reserved in advance. Call the Yoho–Burgess Shale Research Foundation (250–343–6480 or toll-free in America 1–800–343–3006). You must be in good health as the walk is strenuous and involves some steep climbs. Collection of fossils is absolutely forbidden and the area is closely monitored. There is a hiking path (Burgess Pass–Yoho Pass) that passes quite close to the quarries, and in good weather provides a superb day in spectacular scenery. For further information apply to the Superintendent, Yoho National Park, PO Box 99, Field, British Columbia, Canada. Tel. 250–343–6324. Further developments in the Park to assist visitors appreciate the geology and in particular the Burgess Shale are under active consideration.
2. Chengjiang, China. The localities are near Kunming in Yunnan province, China. Several research groups from Nanjing, Xi'an, and Kunming are actively excavating. Geological field trips for professional scientists are run quite frequently at the localities.
3. Sirius Passet, Greenland. This locality is extremely remote and requires a fully equipped expedition. Research work requires a permit and permission from the military authorities.

Index